建筑立场系列丛书 No.45

C3

建筑的文化意象
Cultural Image of Architecture

中文版
(韩语版第361期)

韩国C3出版公社 | 编

时真妹　曹硕　张琳娜　周一　蒋丽 | 译

大连理工大学出版社

建筑立场系列丛书 No.45

4
004 白树塔
　　_Sou Fujimoto Architects + Nicolas Laisné Associés + Manal Rachdi Oxo Architects
008 柏拉图式住宅塔楼_Tammo Prinz Architects
012 Archivo设计品陈列楼_Zeller & Moye + FR-EE

16 建筑的文化意象

016 建筑的文化意象_Silvio Carta

18 文化建筑新方向

018 文化建筑新方向_Douglas Murphy
024 De Nieuwe Kolk文化建筑群_De Zwarte Hond
038 萨拉戈萨艺术馆_Estudio Carme Pinós
050 Eemhuis文化中心_Neutelings Riedijk Architecten
062 21世纪国家电影档案馆_Rojkind Arquitectos
072 大东文化艺术中心_MAYU architects+ + de Architekten Cie
084 迈阿密佩雷斯艺术博物馆_Herzog & de Meuron
100 路易斯安那州立博物馆和体育名人堂_Trahan Architects
112 波兰犹太人历史博物馆_Lahdelma & Mahlamäki Architects
124 拉梅骨多功能亭台_Barbosa & Guimarães Arquitectos

132 曾经的工业建筑向文化建筑的转变

132 从工业到文化，人员流动取代商品流通
　　_Tom van Malderen
138 建在旧储气罐内的多功能大厅_AP Atelier
154 C-Mine建筑_51N4E
166 胶印厂的改造_Origin Architect
176 人人剧院_Haworth Tompkins Architects

188 建筑师索引

4
004 White Tree Tower
_ Sou Fujimoto Architects + Nicolas Laisné Associés + Manal Rachdi Oxo Architects
008 Platonian Housing Tower _ Tammo Prinz Architects
012 Archivo Design Collection Tower _ Zeller & Moye + FR-EE

16 Cultural Image of Architecture

016 *Cultural Image of Architecture _ Silvio Carta*

18 New Directions in Cultural Buildings

018 *New Directions in Cultural Buildings _ Douglas Murphy*
024 Culture Complex De Nieuwe Kolk _ De Zwarte Hond
038 CaixaForum Zaragoza _ Estudio Carme Pinós
050 The Eemhuis Cultural Center _ Neutelings Riedijk Architecten
062 21C National Film Archive _ Rojkind Arquitectos
072 Dadong Arts Center _ MAYU architects+ + de Architekten Cie
084 Pérez Art Museum Miami _ Herzog & de Meuron
100 Louisiana State Museum and Sports Hall of Fame _ Trahan Architects
112 The Museum of the History of Polish Jews _ Lahdelma & Mahlamäki Architects
124 Lamego Multipurpose Pavilion _ Barbosa & Guimarães Arquitectos

132 Cultural Shift from bygone Industry

132 *From Industrial to Cultural, Substituting a Flow of Goods with a Flow of People
_ Tom van Malderen*
138 Multipurpose Hall Fitted in Former Gasholder _ AP Atelier
154 C-Mine _ 51N4E
166 Rebirth of the Offset Printing Factory _ Origin Architect
176 The Everyman Theater _ Haworth Tompkins Architects

188 Index

城市建筑综合体 Urban Complex Tower

白树塔 _Sou Fujimoto Architects + Nicolas Laisné Associés + Manal Rachdi Oxo Architects

跨越多个学科的设计团队藤本壮介建筑事务所、尼古拉斯·莱斯内协会和Manal Rachdi Oxo建筑师事务所,以及开发商Promeo Patrimoine和Evolis促进会,赢得了蒙彼利埃的第二届"21世纪疯狂建筑比赛"。

"21世纪疯狂建筑比赛"是日本和地中海地区建筑风格碰撞的产物。这种跨文化努力的结果呈现出一个现代化的蒙彼利埃。同时也是两代建筑师交流的结果,包括代表日本当前建筑水准的藤本壮介公司以及代表法国年轻一代的Manal Rachdi Oxo建筑师事务所和尼古拉斯·莱斯内协会。其他的公司也被带进这一冒险的事业:包括蒙彼利埃的开发商Promeo Patrimoine+Evolis促进会和当地的股东,这些股东将会确保整个标志性项目的实施,所有人员将促进整个地区的成功。

新的多功能塔称为Arbre Blanc(白树),设有住房、一个餐厅、一个艺术画廊、办公室、全景酒吧和公共区。从项目的概念阶段来看,建筑师的大量灵感来自蒙彼利埃户外生活的传统。建筑坐落在市中心和新开发的玛丽安妮港口和奥德修斯港口地区之间,即新老蒙彼利埃地区的交界处,具有战略性意义。

项目也位于几个街区的路口:Lez河、汽车道和"蒙彼利埃入市税征收处"或政府赠与地的银行周围的人行横道/环路。

这个项目以隆重的姿态展开,来对Lez河沿岸的景观公园进行延伸,并延伸了克里斯托弗·克罗姆地区的长度。东面沿着环状交叉路口的边缘呈现迂回的形式,而Lez河一侧的西面则是凸出的形状,提供非常广阔的全景视角。曲面主要有两个功能:这部分立面既可以提供绝佳的露天景观也不会阻碍邻居的视线。

这座建筑尊重并且融合到周边环境中,展示出自己独特的、额外的魅力。拱形结构像一对翅膀拥抱着向下蜿蜒至庞贝大道的Lez河的轮廓,而"白树"就像经过水和风的雕刻,刻意以自然的风貌屹立于此。

这座建筑完美地模仿了树的形状,成为周围环境的一部分,同时又提供了更多遮阴的地方,使其更像是一处环境。

尽管其名字为"白树",但这里绝不是一个象牙塔。就像融合到都市歌曲中的一个音符,这座建筑注定成为蒙彼利埃市的每个人心中的高层公共建筑。这座蒙彼利埃建筑群中最疯狂的高层建筑正成为城市的焦点,也成为一个地标,成为一个灯塔以及都市夜晚上空的启明星。

White Tree Tower

The multidisciplinary team including Sou Fujimoto Architects, Nicolas Laisné Associés, Manal Rachdi Oxo Architects and the developers Promeo Patrimoine + Evolis Promotions, have won the competition for the construction of the second "Architectural Folie of the XXIth Century" in Montpellier. This "Architectural Folie of the XXIth Century" was mainly the brainchild of an encounter between Japan and the Mediterranean. The cross-cultural endeavour embodies modern Montpellier. It is also an interchange between two generations of architects, with Japanese firm Sou Fujimoto at the state of its art and the young French generation represented by Manal Rachdi Oxo Architects and Nicolas Laisné Associés. Other firms were brought in to take part in this one-of-a-kind venture: Montpellier developers Promeo Patrimoine and Evolis Promotion, local stakeholders who will ensure this iconic project will promote the success for the entire region.

The new multipurpose tower called "Arbre Blanc" which means white tree is designed for housing, a restaurant, an art gallery, offices, a bar with a panoramic

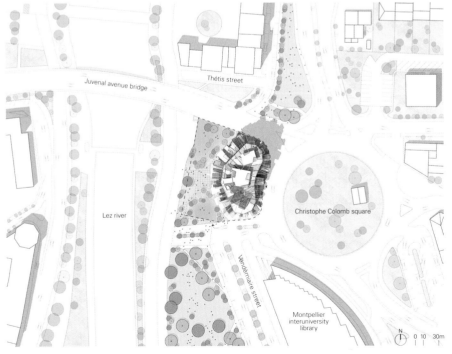

连续的景观
continuous landscape

融合&城市移动
integration & urban movement

尊重邻里建筑的视野
respect for the views of the neighboring building

面向Lez河开放
open to the Lez River

聚光和挡风
catch the light & break the wind

居住在室外，享受风景
live outside and enjoy the views

view and a common area. From the project's concept phase, the architects were heavily inspired by Montpellier's tradition of outdoor living. The tower is strategically located between the city centre and the newly developed districts of Port Marianne and Odysseum, midway between the "old" and the new Montpellier.

It is also situated at the crossroads of several thoroughfares: the Lez River, the motorway and the pedestrian/cycling path along the banks of the "octroi de Montpellier", or land grant.

The project will kick off with a grand gesture to extend a landscaped park along the Lez and stretch out the length of Christophe Colomb Place. The eastern face curves along the edge of the roundabout while the western side on the Lez is convex to create the widest panorama possible. The curvature serves two purposes because this part of the facade offers the best exposure and viewpoint but does not block the view for neighbouring residences.

The building is sited to meld with and defer to its surrounding environment, yet gives it

just the right added flair. Arching like a pair of wings hugging the contours of the Lez River down to Pompignane Avenue, Arbre Blanc was intentioned as a natural form that was carved out or sculpted over time by water or wind. It perfectly mimics a tree reshaping itself to grow into its environment yet simultaneously enhancing it by offering much-needed shade.

Despite the name "white tree," this is by no means an ivory tower. Like a beat integral to the urban song, the building is destined as a public high-rise built for every soul in Montpellier. This tallest "Folie" in Montpellier's architectural arsenal is becoming the city's focal point, a landmark that serves as a lighthouse or guiding star at night amid the regional urban skyline.

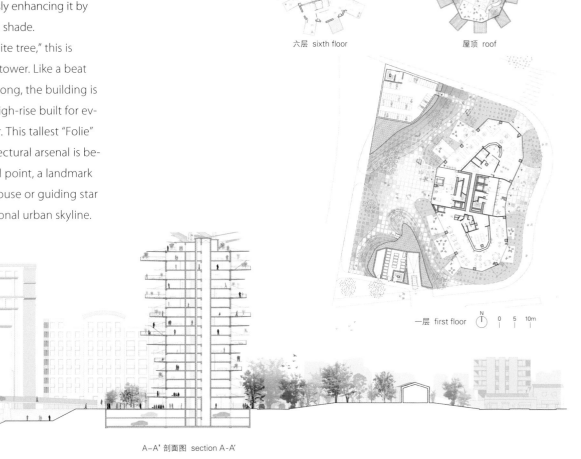

六层 sixth floor　屋顶 roof

一层 first floor

A-A' 剖面图 section A-A'

城市建筑综合体 Urban Complex Tower

柏拉图式住宅塔楼_Tammo Prinz Architects

德国的建筑办公室塔姆·普林茨建筑师事务所设计了一座住宅摩天大楼,利用一堆立方体模块和十二面体堆叠建成。建筑师利用彼此镶嵌在一起的三维模型的组合,重塑了位于秘鲁利马的一处场地。

Peruan建筑规范的使用使斜坡式的外围护结构呈现最大化,项目体量非常大。因此,设计引入了外部的下沉广场,与一处开放且公共的接待空间相连接,作为可能的举行活动的场所。这样,生活空间可以移置靠近一层,仍旧可以通过下沉广场所产生的缝隙空间来提供足够的私密性。

下沉广场由圆形场地式的阶梯所包围,这些阶梯由五角形的岩石组成,以和建筑相契合。在几个位置处,岩石(座位)在任意高度被同样形状但不同颜色的石质阶梯穿过。所有的公共空间都位于地下,地下区通过连向接待处的大型空间,来吸收自然光进行照明。

五个柏拉图住宅主体中的两个(立方体和正十二面体)以其自身的特征而被选中为其特色形状,它们彼此搭配得天衣无缝。通过这样所塑造的空间具备两个特质:一是它可以被当成内部空间,此外还可以作为潜在的附属空间。

首先,完美的正方形半面很容易成就居住功能,其次,也可以融入到其他空间当中成为其中的一部分或者作为外部空间。正十二面体作为巨大的混凝土结构,其轮廓让内部的空间非常自由。建筑的外表多是不光滑的混凝土结构,六个正十二面体一起创造了内部的多余空间,形成星星的形状,且完美地成为了建筑的基础结构。

另外,星星的轮廓和流入正十二面体的立方体相呼应,并且将正十二面体空间内部的立方体转为外部空间,覆盖着星体结构。

Platonian Housing Tower

German Architectural office Tammo Prinz Architects designed a residential skyscraper built from a stack of modular cubes and dodecahedrons. They have been redeveloping a site in Lima, Peru, using a

combination of three demensional shapes that tessellate with one another.

The sloped maximum envelope was given by Peruan building codes. The program was massive. Therefore, the outside sunken plaza was introduced, connected to an open and public reception space, serving together as possible event space. In this way, the living function could reach down close to the ground floor, still providing enough privacy through the gap generated by the sunken plaza.

The sunken plaza is bordered by amphitheater like steps, formed from Pentagon-shaped rocks(corresponding to the building). The random heights of the rocks/seats are traversed at several positions by steps of the same shape in different colored stones. All common spaces are positioned underground exposed by natural light through generous voids into the reception space. Two of the five Platonian Bodies(Cube and Dodekaeder) are chosen for their characteristics, that they perfectly fit into each other. By this generating a space clearly defines two qualities – one to be taken as inside space, the other as the potential additions.

The first, a perfect square is easy to handle the living functions, the second can be either added to those or be used as outside space. The outlines of the Dodekaeder serve as a massive concrete structure, giving total freedom to the inside. The building's appearance is mostly defined by its brute concrete structure. 6 Dodekaeders moved together generate a left over space inside, that shapes a star and perfectly serves the Dodekaeders of

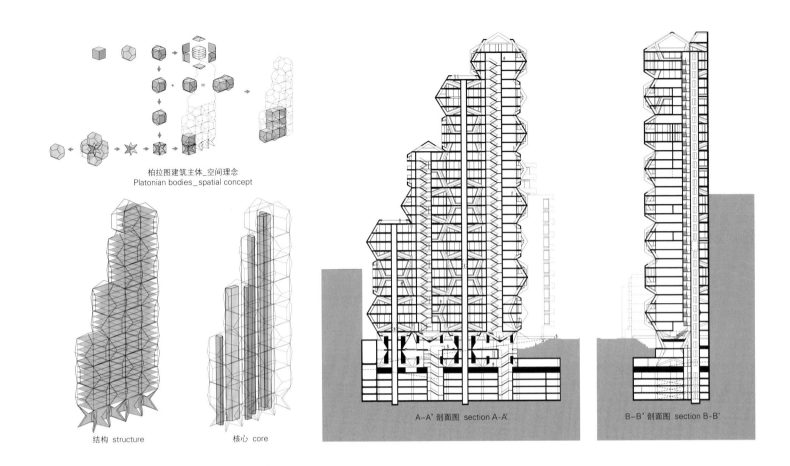

柏拉图建筑主体_空间理念
Platonian bodies_spatial concept

结构 structure

核心 core

A-A' 剖面图 section A-A'

B-B' 剖面图 section B-B'

the tower as carrying base structure. In addition, the boundaries of the star correspond exactly with cube moved into the Dodekaeder, converting the cube from inside space within the Dodekaeder, into outside space, coating the star structure.

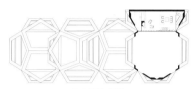

三十三层_阁楼
33rd floor_penthouse

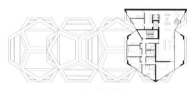

三十二层_阁楼
32nd floor_penthouse

十五层_公寓
15th floor_apartments

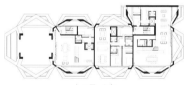

十四层_公寓
14th floor_apartments

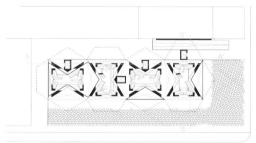

一层 first floor

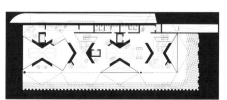

地下三层 third floor below ground

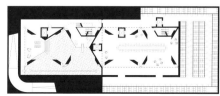

地下二层 second floor below ground

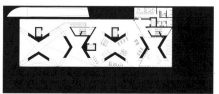

地下一层 first floor below ground

Archivo设计品陈列楼 _Zeller & Moye + FR-EE

拉美的第一个设计品陈列新家由Zeller&Moye和FR-EE事务所合作建造,为墨西哥城市当代繁荣的设计文化增添了色彩,让这座建筑在城市中的形象更加鲜明。这个项目计划于2014年年末破土动工。

"两年后,建造一个全新的地面设施,并且在这个设施中创造和设计新的展览品的理念着实令人兴奋。"Archivo设计品陈列楼的主管瑞吉娜·坡左说到。于2012年启动,以促进和展示20世纪早期至现在的工业设计的精华为己任,Archivo设计品陈列楼已经日渐成为墨西哥城学习和体验设计的中心。该建筑收集了1300个展览品,数量超过了附近现有的画廊,以及临近的塔库巴亚地区的现代主义建筑师刘易斯·巴拉的工作室的展品数量,并且致力于进一步完善展览和各项操作。这个项目预计于2014年末开始建设。

新的Archivo设计品陈列楼位于墨西哥城市中心的一处场地,被美丽的城市花园所包围,给首都这片未被发现的地区带来了生命和可再生的能量。风格各异且透明的长廊除了迎接访问者到里面观看专有不变的设计项目展览之外,还能使他们享受不同的功能和活动。建筑师优先设计了社会活动、谈话、商业活动的空间,以创造流动的氛围,有利于对话和重要的文化交流。

3000m²的建筑被设计为6层的原始骨架,对奇异的外部空间开放。这座建筑的结构包括一个垂直的核心和水平楼板,楼板向花园内部延伸,创造出非凡的内外空间的组合。一个螺旋形的楼梯沿着周边或延伸或缩进,引领着访问者从地下,通过展览的内部和外部区域,一直来到公共屋顶平台,来欣赏整座城市的风景。楼梯作为外部的展示空间或者非正式的会议和休息区域,很适合墨西哥常年温暖的气候。

玻璃立面从嵌板边缘便开始采用退后设置，以形成隐蔽的区域，提供一定的隐私性，并且与简明的结构相得益彰。同时灵活的边缘处还设置了更多的公共空间。

围绕建筑形成的公共生活圈是整个项目的一部分。多功能区域包括工作间、舞蹈教室和社交场所，还有围绕Archivo的城市花园室外区域，都会成为当地社区人民和来墨西哥城参观的人们的目的地。

Archivo Design Collection Tower

The new home for Latin America's first design collection designed by Zeller&Moye in collaboration with FR-EE adds to Mexico City's flourishing contemporary design culture and gives the collection a stronger presence in the city. It is scheduled to break ground in late 2014.

"After two years, the thought of a new ground-up facility in which to create and design new shows is thrilling." says Archivo director Regina Pozo. Established in 2012, with the mission to promote and exhibit the best of industrial design from early 20th century to present, Archivo has become the go-to hub for learning and experiencing design in Mexico City. Archivo's collection of 1,300 objects is outgrowing the existing gallery space next to the house and studio of modernist architect Luis Barragan in the neighborhood of Tacubaya and seeks to further consolidate its exhibitions and operations in the new building. The project is expected to start

construction by the end of 2014. Located on a site in the heart of Mexico City surrounded by luscious jungle-like gardens, the new Archivo brings life and regenerative energy to an undiscovered part of the capital. A diverse and transparent gallery space welcomes the visitors inside to enjoy a variety of functions and activities beyond the permanent collection of exclusive design items. Spaces for social events, talks and commercial use have been prioritized to create a more dynamic atmosphere, facilitating dialogue and critical cultural exchange.

The 3,000m² building is designed as a raw exoskeleton of six levels that opens up to the exotic surroundings. The structure of the building consists of a vertical core and horizontal floor plates that branch out into the garden, creating an unusual mix of indoor and outdoor spaces. A spiraling staircase expands and contracts along the

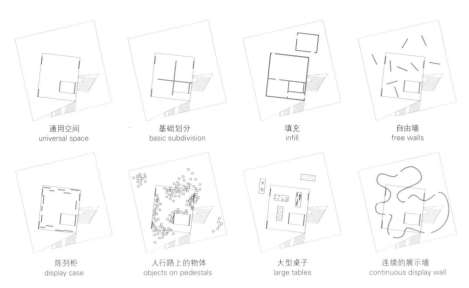

perimeter, leading the visitors efficiently from ground floor, through the exhibitions inside and outside, all the way up to the public roof terrace to enjoy spectacular views of the city. The staircase serves as an outdoor exhibition space or simply as an informal meeting and resting area, ideal for Mexico's year-long moderate climate. The clean structure is completed by glazed facades set back from the slab edge to provide shade and privacy, whilst the more public functions are placed along the active edges.

A spectrum of communal life forms around the building are an integral part of the project. Multi-functional spaces for workshops, dance classes and socializing, as well as outdoor areas for urban gardening surrounding Archivo will serve as a destination for the local community and visitors of Mexico City.

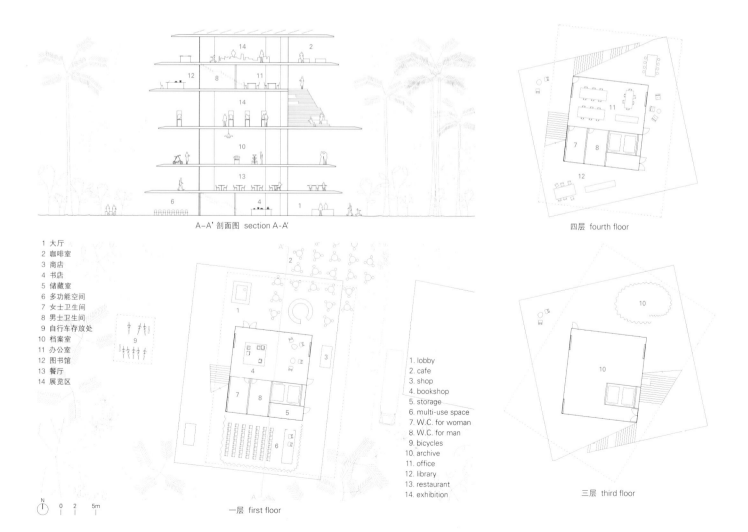

1 大厅
2 咖啡室
3 商店
4 书店
5 储藏室
6 多功能空间
7 女士卫生间
8 男士卫生间
9 自行车存放处
10 档案室
11 办公室
12 图书馆
13 餐厅
14 展览区

1. lobby
2. cafe
3. shop
4. bookshop
5. storage
6. multi-use space
7. W.C. for woman
8. W.C. for man
9. bicycles
10. archive
11. office
12. library
13. restaurant
14. exhibition

A-A' 剖面图 section A-A'
四层 fourth floor
一层 first floor
三层 third floor

建筑的文化意象
Cultural Image

"文化(culture)"一词源于拉丁语动词colere(居住、培养、常到、实践、倾向、注意、警惕的意思),并包含了一系列与培养和土地耕种有关的术语[1]。这是一种比喻性的说法,文化是指"通过教育的方式培养"[2],其中通过有组织的、耐心的、勤奋的工作使泥土中的作物日益生长,是比喻借助于教育获得知识的增长。如今,文化的一般概念与其农业的意思分离,是与"依靠学习能力和知识的传承而形成的一种人类知识、信仰和行为的模式"相关联的,同时也与"一个种族、宗教或社会团体的传统信念、社会形态以及物质特性"息息相关[3]。

从词源学意义上来简单地解读"文化"这个词,我们注意到个人的修养与其知识水平及其他个体的关系是紧密相连的。人类文化起源于过去,在现代一个有组织的社会中开展,并将目光投向未来。因此,教育的主要特征之一即是依赖于对过往知识的保留、传播和交流的。转移的过程可以采用两种途径:有意识的——即当有人掌握着特定的知识并将其传播给特定的群体的时候,以及无意识的——即当学习者观察、解读和思考这个世界的时候。

如果世界上的任何实际事物中都蕴含文化的元素,那么经历一个解读和加工的过程,我们可以这样说,世间万物都潜藏着一些可以转化成为文化元素的知识。

如果说文化深藏于一切事物中,那么,建筑与它的关系是怎样的呢?一方面,我们能够感知到建筑的"作者们"的明确意图,然而另一方面,我们发现建筑能够间接地讲述与自己有关的许多事情。它的表达极为清晰,比如,一座建筑是在怎样的条件下、被特定的时间和地点付诸实践的,它包括了一长串的因素。总而言之,一个完成的作品是其起源于复杂特性的物理表现形式。一旦

The word "culture" stems from the Latin verb colere(to inhabit, cultivate, frequent, practice, tend, take care of, guard), and comprises a series of other terms related to the cultivation and tillage of the land[1]. Figuratively, culture has come to refer to "cultivation through education",[2] where the growing of goods from the soil through organised, patient, hard work is compared to the increase of knowledge by means of education. Apart from its agricultural meaning, the general notion of culture now relates to "a pattern of human knowledge, belief, and behavior that depends upon the capacity for learning and transmitting knowledge to succeeding generations" and to "the customary beliefs, social forms, and material traits of a racial, religious, or social group".[3]

From this brief reading of the etymology of the word "culture", we note that the cultivation of the individual is deeply linked to previous knowledge, as well as to relationships with other individuals. The culture of people stems from the past, operates in the present in an organised society, and casts directions for the future. One of the main characteristics of education thus hinges on the preservation, diffusion, and communication of past knowledge. This process of transfer can happen two ways: consciously – when someone holding particular knowledge broadcasts it to specific groups of people – and unintentionally – when learners observe, interpret and think about the world.

If elements of culture can be found in practically everything in the world, undergoing a process of interpretation and elaboration, we could say that potentially everything carries some knowledge that can be turned into a cultural element.

If culture, then, is embedded in virtually everything, how does architecture relate to it? On one side we may perceive the clear intention of the authors of architecture, while on the other side we find a building can indirectly say a great deal about itself. It expresses quite clearly, for instance, the conditions under which architecture has been realised in its specific time and place, including a long list of factors. In sum, a built artefact is the physical representation of the intricacy of idiosyncrasies from which that architecture originated. Once completed, that architecture is the picture of that culture frozen in the unique moment of its construction.

# of Architecture

完成，建筑就成为了一幅文化图片，将时间凝结在了它竣工的那一特定时刻。

后面的两个章节从广义上探讨了建筑与其文化框架的关系，在"文化建筑新方向"章节中，道格拉斯·墨菲详细说明了过去几十年中标志性的"大型文化建筑"已经转变成为更为复杂的类型。这样的文化建筑"可能因其多功能性而带有种种不同的功能，建造的建筑较之前会在其经济和社会属性上更加多样化"。本节内容所包括的各种各样的项目表明诱人的、强有力的并且引人注目的都市元素正在向"在其使用及提供的公共空间方面更加慷慨、更加复杂的文化建筑"的方向平缓过渡。

在"曾经的工业建筑向文化建筑的转变"章节中，Tom Van Malderen研究了文章"从工业到文化"以及随之发生的"人员流动取代商品流通"的过程。文章对废弃的工业基础设施和旧区的改造项目中（一座建筑由工业产品转变成为文化产品）的新文化活动之间的关系进行了分析。而且，人们会逐渐意识到每一座建筑都承载了大量的文化元素，它们来自于建筑的过去以及居住和使用这座建筑的人们，并成为了观察新项目在改造原有建筑的过程中是如何处理其历史传统的关键所在。在"曾经的工业建筑向文化建筑的转变"章节中对提到的项目进行分析时也对建筑肌理、位置、场所的文化意义或是现代与过去的碰撞这样几点问题进行了探讨。

本期为人们观察到的文化项目提供了两种不同但却互补的角度。尽管它们的形式和背景都相去甚远，但这些建筑都呈现出了一种特有的文化成分，能够洋洋洒洒地向人们讲述自己独有的故事。它赋予了人们火眼金睛，使人们能够通过建筑的立面看清其本质甚至是更深远的东西。

The two following chapters explore the relationship between architecture and its cultural framework in its broader definition. In "New Directions in Cultural Buildings" Douglas Murphy elaborates on the shift from the "large cultural building" which characterised the last few decades to a more sophisticated typology in which cultural buildings "might be cross-programmed with a variety of different functions, creating architecture more economically and socially diverse than before". The various projects included in this section illustrate the gradual turn of cultural buildings from seductive, powerful and eye-catching urban elements to "cultural buildings which are becoming more generous and more sophisticated in their use and provision of public space".

In "Cultural Shift from bygone Industry", Tom Van Malderen investigates the passage "from industrial to cultural" and the consequent "substitution of a flow of goods with a flow of people". The chapter analyses the relationship between abandoned industrial infrastructures and new cultural activities in revitalisation projects where "a building shifts from industrial production to cultural production". Moreover, as one becomes conscious that every building carries a cloud of cultural elements from its past and from the people who inhabited and used it, it becomes crucial to observe how the new project transforms the existing architecture while dealing with its heritage. In "Cultural Shift from bygone Industry" several questions such as "the building fabric, the location, the cultural significance of the place or the encounter between the present and the past" are discussed in the analysis of the presented projects.

This issue offers two different, yet complementary angles from which we can observe the cultural projects presented. Although in different manners and in contexts remote from one another, these buildings all evince a peculiar cultural component which can be quite eloquent concerning each building's unique story. It is down to a trained eye to see through and beyond the facade. Silvio Carta

1 The online version of the Merriam-Webster dictionary holds that the term originates from the Anglo-French term, indicating cultivated land, cultivation, from Latin "cultura", from "cultus". http://www.merriam-webster.com/dictionary/culture retrieved August, 13, 2014.
2. Cf. http://www.etymonline.com/index.php?term=culture retrieved August, 13, 2014.
3. Both definitions are from http://www.merriam-webster.com/dictionary/culture retrieved August, 13, 2014.

文化建筑新方向
New Directions in

近年的全球金融危机给建筑带来了一种新的严肃气息。纵观全球,众多建于本世纪初经济热潮的文化建筑以失败告终,只留下建筑设计的宏伟外观,无法实现雄心勃勃的项目书要求。由于受到严厉的政治修辞的影响,出现了一种新的约束。我们可能处于一个连接点上:在阿布扎比,这个时代一些最宏伟的建筑仍未完工,包括由老一代巨星设计的标志性文化建筑。当初十年完工的计划让人感觉,这可能是标志性时代的最后一些主要建筑物。但现在看来这些建筑思想似乎有点声名狼藉。

文化建筑过去被理解为城市图标,而建筑师打破这种局限的尝试在近年的新型文化建筑中得到了体现。尽管仍常常在标志性建筑的总体框架下操作,但非常明确的是,回归语境的复杂性和对城市公共空间的关注变得比以前更加突出。我们审视了这些设计,不免要问:有没有可能上个十年最严重的建筑过剩现象如今已经远离我们?"酋长国风格"的建筑有能力在城市里发挥作用吗?

Since the global financial crisis of recent years began, it seemed that a new sobriety was appearing in architecture. Across the world, a large number of cultural buildings, commissioned during the economic fever of the early 2000s, failed, leaving architecturally spectacular buildings completely incapable of fulfilling their over-ambitious briefs. Since then, a new restraint has appeared, influenced by the politics rhetoric of austerity. Today we are perhaps at a junction point: in Abu Dhabi, some of the most ambitious architecture of the era is currently under construction, including signature cultural buildings by all the superstars of the last generation. Due to be completed before the end of the decade, there is a sense that these might be the last major buildings of the iconic era, a by-now somewhat discredited architectural ideology.

A number of new cultural buildings of recent years show how architects are attempting to move beyond the restrictions of the cultural building understood as the urban icon. Though still often operating within the general parameters of iconic architecture, it is clear that a return to contextual sophistication and a commitment to the public space of the city are becoming more prominent than before. By examining some of these designs, we can begin to ask: is it possible that the worst architectural excesses of the last decade are now behind us? Is the "Emirate style" of architecture capable of acting urbanistically after all?

De Nieuwe Kolk文化建筑群_Culture Complex De Nieuwe Kolk/De Zwarte Hond
萨拉戈萨艺术馆_CaixaForum Zaragoza/Estudio Carme Pinós
Eemhuis文化中心_The Eemhuis Cultural Center/Neutelings Riedijk Architecten
21世纪国家电影档案馆_21C National Film Archive/Rojkind Arquitectos
大东文化艺术中心_Dadong Arts Center/MAYU architects++de Architekten Cie
迈阿密佩雷斯艺术博物馆_Pérez Art Museum Miami/Herzog & de Meuron
路易斯安那州立博物馆和体育名人堂_Louisiana State Museum and Sports Hall of Fame/Trahan Architects
波兰犹太人历史博物馆_The Museum of the History of Polish Jews/Lahdelma & Mahlamäki Architects
拉梅骨多功能亭台_Lamego Multipurpose Pavilion/Barbosa & Guimarães Arquitectos
文化建筑新方向_New Directions in Cultural Buildings/Douglas Murphy

Cultural Buildings

在建筑世界里，几乎整整一代人都在无助地痴迷于一种特殊的形态：大型的文化建筑。由弗兰克·盖里设计的、于20世纪80年代末接受委托，1996开放的毕尔巴鄂古根海姆博物馆，由于大肆宣传而获得成功。从此，世界各地的城市和客户开始迷恋世界著名建筑师设计的宏伟建筑，来协助所在城市在全球投资竞争中和环境命运的再生中吸引游客。世界各地花钱无数去追逐标志性建筑的积极影响，上一代中许多最重要的建筑就是在这个过程建成的。

世界形势对建筑师和客户都是利好消息，甚至在金融危机之前就有评论家做出了评论。2005英国作家乔纳森·米德提出"视口"的建筑形式，描述了在诱惑，奢侈而又常为肤浅的建筑创造中，宏伟建筑和城市的复杂性和微妙性是如何迷失的。米德认为，这种新一代"伪现代"建筑的功能无异于推进房地产迅猛发展和促变城市为新型富有阶层的催化剂，许多城市特别是欧洲和北美经济发达的地区确实如此。

诚然，经济原因让文化引导型的再生设计似乎永远名誉扫地，很多的故事也证明了建筑界在经济崩溃后几年内逐渐没落。有些项目很成功，如古根海姆博物馆，但也有被经济危机击垮的项目。最可悲的就是西班牙北部省区的加利西亚文化城，人们认为这个超复杂的建筑群是彼得·艾森曼建成的人造景观，但由于极端浪费可能永远只完成一半。最近，历史学家约瑟夫·里克沃特针对21世纪的标志建筑创造了"酋长国风格"一词，他认为"建筑整体实际上是一个商标，会呈现出与广告的一种矛盾关系。"随着古根海姆博物馆迥异的前哨，卢浮宫、其他建筑师如弗兰克·盖里，扎哈·哈迪德和让·努维尔在阿联酋阿布扎比的设计，受到了客户最热情和最包容的欢迎，但也可能意味着这一时期的建筑历史即将终结，因为全球动荡和气候变化的重大危机使现状变得陌生和更加不确定，在这种情境下，世界的优先权发生了变化。

这并不是说在未来就不会有人去建造文化建筑，但实际上，从客户

For almost a generation now, the world of architecture seemed helplessly obsessed with one particular typology: the large cultural building. From the obvious and much-hyped success of Frank Gehry's Guggenheim Museum in Bilbao, commissioned at the end of the 1980s and opening in 1996, cities and clients worldwide became obsessed with commissioning spectacular cultural buildings designed by globally renowned architects, which would be tasked with attracting tourists, assisting their host cities in the global fight for investment, and often the complete regeneration of the fortunes of their environment. Huge amounts of money were spent across the globe chasing these positive effects of signature architecture, and many of the most important buildings of this last generation were created through this process.

But while this world was good news for the architects, and frequently good news for the clients as well, even before the financial crisis there were critics. In 2005 British writer Jonathan Meades coined the term "sight-bite" for this form of architecture, describing how the sophistication and subtlety of great buildings and cities were being lost through the creation of such seductive, extravagant but frequently shallow architecture. In Meades' argument, this new generation of "pseudomodern" buildings was frequently functioning as little more than catalysts for aggressive real estate developments and the transformation of cities for a new rich urban class, a statement which rings true for a great many cities, especially those of the developed economies in Europe and North America.

Of course, many of the narratives justifying this world of architecture evaporated in the years after the crash, with the economic rationale for culture-led regeneration seemingly discredited forever. For every successful project like the Guggenheim Bilbao, there were others which were caught out by the crash, perhaps none more pathetic than the City of Culture of Galicia, in a nearby province of northern Spain, a hyper-complex group of buildings conceived as an artificial landscape by Peter Eisenman, which due to extreme profligacy will probably forever remain only half-finished. More recently, the historian Joseph Rykwert has coined the term "Emirate Style" for this new 21st world of signature architecture, in which "buildings assume an ambivalent relation to advertising since their entire bulk is in fact a trademark". With various outposts of the Guggenheim, the Louvre and others designed by the likes of Frank Gehry, Zaha Hadid, Jean Nouvel in Abu Dhabi, the UAE are where this kind of architecture has found its most enthusiastic and accommodating clients, but also perhaps where this period of architectural history will come to a close, as the priorities of the

的期望到设计方法,我们已经看到处理方式的变化。尽管文化建筑作为城市发展的重心仍然有规律地进行着,我们可以看到建筑物以各种方式更智能地连接到环境中。可能被配有多种交叉性的功能,比起以前,它们创造出更多的经济性和社会结构的多样性,或者试图创造一个更敏感的公共空间形式。在某些情况下,这些新的文化建筑可能试图变得更像公园的亭台,回归百年历史之前的形态策略,同时也展现出了新的和更复杂的态度,来对待景观以及建筑和自然形式的关系。一些新的欧洲文化建筑体现出文化建筑在使用和提供公共空间方面变得更大气、更复杂。由Barbosa&Guimarães建筑事务所负责的拉梅骨多功能亭台,为葡萄牙小镇提供了多用途体育馆和礼堂,但项目进展地异常艰难,因为建筑师把建筑的一半埋入了该地区的丘陵地貌,在这个过程中创造出一个新的高于建筑物的地面。一连串的大型楼梯提供了广阔空间(空间可以作为圆形剧场),令边上的市场广场得到凸显。屋顶表面构成一个新的广场,吊着各种各样的灯,照看下面,最高处如皇冠般地安置了一个多用途考顿钢复合亭。该建筑相当内敛,没有侵占新的公共空间。

由Neutelings Riedijk建筑师事务所设计的全新的Eemhuis文化中心,是一座更具形式表现力的建筑,它持续了标志性建筑时代的一些过激行为,但与公共空间的互动更强。这是一个典型的混搭式荷兰作品,在展览厅顶部建成了一所新的艺术学校,而在较低的空间安置了一座公共图书馆。艺术学校的特征为一系列巨梁建筑,浮在玻璃幕墙上面,表面为粗糙的、布满装饰的钢板。建筑物的下部覆盖着复古式的窄砖,更紧密地连接到周围开放的公共广场。建筑师考虑将广场直接延续进大楼,在里面,人们进入展览大厅后向下走,进入图书馆,这是一处置于高处露台上的宽阔的内部空间。对一栋这样的大楼,建筑师关注的不是创作一个非凡的单品和如何表达建筑焦点,而是更加注重墙内的小镇民众生活,使建筑更紧密地融入当地居民的生活。

在荷兰的其他地方,由De Zwarte Hond设计的De Nieuwe Kolk文化建筑群,在现代的框架上采用了类似复古荷兰砖的方法,创造了一座公共建筑,包括剧院、电影院、图书馆和私人住宅。它坐落在一座新建的公共广场的边缘,也试图将公共领域直接带入建筑内部。它拥有一个巨大

world change in this new, more uncertain context of global insecurity and the first major crises of climate change.

This is not to suggest that nobody will build cultural buildings in the future, but it is true that we can already see changes to the way commissions like these are handled, from client expectations through to design approach. While cultural buildings as centerpieces of urban developments are still regularly commissioned, we can see various ways in which the buildings are more intelligently connected into their surroundings. They might be cross-programmed with a variety of different functions, creating architecture more economically and socially diverse than before, or they might attempt to create a more responsive form of public space than we have become used to. In some cases, these new cultural buildings might attempt to become more like pavilions in the park, returning to a typological strategy more than a century old, but also showing new, perhaps more sophisticated attitudes towards landscape and the relationship between a building and the natural form.

A number of new European cultural buildings show ways in which cultural buildings are becoming more generous and more sophisticated in their use and provision of public space. Lamego Multipurpose Pavilion by Barbosa & Guimarães Arquitectos, provides its small Portuguese town with multipurpose sports halls and auditorium. But the program is made to work far harder than this, as the architects have semi-buried the building into the hilly landscape of the region, in the process creating a new ground surface atop the building. A market square to the edge of the site is focused and given prominence by the creation of a series of large staircases providing a grand space which also functions as an amphitheater. The roof surface, which constitutes a new plaza is punctured by a variety of lights which open to the space beneath, and is crowned by a corten steel clad pavilion for multipurpose events. The architecture is fairly restrained, not overwhelming the new civic spaces which the building creates.

The brand new Eemhuis Cultural Center by Neutelings Riedijk Architecten, is a far more formally powerful building, continuing some of the excesses of the era of the icon building, but its interaction with public space is even stronger. It is a typically Dutch exercise in cross-programming, stacking a new art school on top of exhibition halls and a public library at lower levels. The art school at the top is expressed as a series of giant beam buildings apparently floating above a glazed curtain wall, and clad in a rather vulgar studded steel panel. The lower part of the building is clad in a more historically considered narrow brick, which connects the building more tightly to the open public square around it. The architects consider the building to continue this square directly into the building, where the exhibition halls are entered moving downwards, and the library, a vast open internal space stacked on

的公共立面外观。此外，功能区完全区分开来，不同大小的空间堆叠给予不同的外墙处理，进而表达了建筑的多功能性。在Carme Pinós工作室设计的萨拉戈萨艺术馆中，作为休闲和文化场所，公园的使用及其意义得到了提升和重新解释。两个大展厅和后勤部门、餐厅、礼堂和商店等被抬离地面，置入一个进深小但结构表现力很强的塔楼中。悬臂支架连接的两个展厅置于相反方向，建筑的痕迹在地面上减到了最少。楼体坐落于其内的公园被延伸到大厅下面，自动扶梯和巧妙放置的窗口形成了流通通道，创造出有别于公园本身的一种连续的效果，让人联想到穿过城市本身的内在体验。

但是，一座新的文化建筑有时不得不应对不合适的场址，例如由Rojkind Arquitectos建筑事务所设计的位于墨西哥城的21世纪国家电影档案馆的扩建案例。由曼努埃尔·罗查建造的20世纪80年代风格的综合楼，其高度重视车辆通行的特点促成了项目的扩建。Rojkind根据项目书，建造了两座新的剧院和额外的功能设施，在建筑周围创造出一种新的景观类型。针对墨西哥市恶劣的气候，他们创造了一种多孔屋顶结构来遮阴和冷却下面的空间，把停车场分开后，此处的外面可以举办各种活动，如用作户外晚间影院。建筑风格非常流畅，也许有点矫饰，但其城市意图得到了考虑并受到欢迎。

在文化建筑周围创造出一个真正的公共空间的规划是值得称赞的，对城市功能也十分关键，但如21世纪国家电影档案馆所证明的那样，在特别热或潮湿的气候中实现这个目标往往是困难的。由张玛龙+陈玉霖建筑师事务所和Cie建筑师事务所设计的大东文化艺术中心，在中国台湾的热带气候内创造了实用的公共空间，进行了大胆和创新的尝试。构成综合楼的剧院、展览中心、图书馆和教育中心被多种开放的区域分离开来。为了使这些设计更讨巧，屋檐水平的位置上建了一个额外的网格结构，插入巨大的半透明漏斗形表面。这些步骤主要有两个作用：利用烟囱效应吸出公共区域的热气，有利于保持凉爽，还可以将雨水的排水通道隐藏起来，保持干燥，同时在雨季期形成水滴窗帘，为游客创造一个奇观，还不被淋湿。

赫尔佐格&德梅隆建筑师事务所在近期建造迈阿密佩雷斯艺术博

terraces opens out above. The architects' attitude to a building like this is less about a remarkable single institution, architecturally expressed as a focal point, but takes more care over attempting to bring the civic life of the town inside its walls, tying the building closer into the lives of the local inhabitants.
Elsewhere in the Netherlands, the Culture Complex De Nieuwe Kolk by De Zwarte Hond, utilizes a similar language of traditional Dutch bricks on a contemporary frame to create a public building containing theater, cinema, library, and private housing. Situated at the edge of a new public square, it also attempts to bring the public realm directly within the building, with a grand public facade to the exterior. Again, the various different programmes are differentiated and expressed through the stacking of different volumes that give different skin treatments, expressing the multiple functions of the building. The use and significance of parks as a location for leisure and culture are enhanced and reinterpreted in the CaixaForum Zaragoza designed by Estudio Carme Pinós. It sees two large exhibition halls and attendant program areas; restaurants, auditorium, shops etc, lifted off the ground plane into a short, structurally expressive tower. The two exhibition halls are dramatically cantilevered in opposite directions, with the building footprint reduced to a minimum at ground level . The park in which the building sits is extended right underneath the halls, and the circulation throughout is set up using escalators and strategically placed windows to create a procession which is minimally differentiated from the park itself, to create an internal experience which is reminiscent of the act of walking through the city itself.
But occasionally a new cultural building has to deal with an unsympathetic site, which is the case for the extension to the 21C National Film Archive by Rojkind Arquitectos in Mexico City. The original 1980s complex, by Manuel Rocha, was highly focused on vehicle access, which gave the main impetus to the extension project. Rojkind approached their brief, which included two new theaters and extra program, as an opportunity to create a new kind of landscape around the building. In the difficult climate of Mexico City they created a perforated roof structure which shades and cools the spaces underneath, which along with the separation of car parking allow for a variety of activities to take place in the exterior, including an outdoor cinema for evening screenings. The architectural style is very slick, perhaps a little pretentious, but its urban intentions are considered and welcoming.
The intention to create a genuine public space around a cultural building is laudable, and vital for its civic function, but as the 21C National Film Archive shows, it is often a difficult thing to achieve in particularly hot or humid climates. Dadong Arts Center by MAYU architects and de Architekten Cie makes a bold and inventive gesture for creating useful public space in the tropical climate of Taiwan, China. The buildings which make up the complex – a

物馆时遇到了相似的挑战。炎热和潮湿的环境与对当代艺术的冷静反思经验毫无关系，所以建筑师必须创新。利用一个大于内部空间边缘的巨大底座，实现更大程度的灵活性：该底座比潜在的洪水的最大高度还高，创造了一个类似密斯式的矮墙。屋顶结构不仅让底座的高度增加了一倍，还给画廊周围带来阴凉的微气候。屋顶下的葡萄树和其他植物慢慢地改变了画廊有些斯巴达式的正规排列，既能进一步改善大气条件，也会最终给博物馆的外观带来一种热带雨林的效果。很明显，这些新的文化建筑虽然体积庞大，壮观宏伟，但涉及了尝试人工景观形式的各种策略，并力图紧密地融合在当地的背景下。我们同时还看到，过去十五年没有对大型文化建筑的气候策略进行调查，但它只会变得更加重要。但自然主义的隐喻在这段时期一直是标志性建筑的重要组成部分，许多建筑被出售时都会被比拟成宝石、珍珠、河流和其他现象。两个最近的文化项目都采用了这种想法，但比起以前，它们更有控制力，也不太体现消费态度了。

不可避免地，Lahdelma&Mahlamäki建筑事务所设计的波兰犹太人历史博物馆和丹尼尔·李博斯金设计的1996年柏林犹太人博物馆形成对比。两座建筑利用说教式的展览，尝试讲解一个更完整的犹太人的故事，让人们知道二战大规模谋杀之前在中欧和东欧他们是一个成规模的少数民族。但李博斯金的建筑，这也是他最好的作品，把建筑本身作为一个高度象征的概念对象，其功能和展览相得益彰，而波兰建筑师的建筑却尝试聚焦几个非常具体的特点。该建筑位于华沙犹太人区起义开始的地方，规划十分简单，采用刻有拉丁文和希伯来文的玻璃幕，但贯穿整座建筑的是一处大型峡谷般的空间，用抛光的喷射混凝土建成，类似沙漠风沙吹成的岩石雕刻或一个干涸的河床。这种空间的背后的理念是文本式的，镌刻的希伯莱语"Yam Suf"，意为"红海的分离"，表示被杀害的犹太人的缺席，也意味着幸存者和他们文化的连续。但这个峡谷并没有控制建筑整体，只是执行了一个象征性的功能，因此整座建筑的功能并未受到影响。虽然讲述的故事并不那么感伤，Trahan建筑事务所设计的路易斯安那州立博物馆和体育名人堂也采用了类似方法。该建筑是一个相当传统的箱形结构，外表是铜色的褶皱，使光可以进入建筑内

theater, an exhibition center, a library and education center – are separated from each other by various open areas. To make these hospitable, an additional gridded structure at eaves level is created, into which large translucent funnel-shaped surfaces are inserted. These work in two primary ways: they utilize stack effect to draw hot air out of the public areas, helping to keep them cool, and they channel rainwater into hidden drains, keeping visitors dry and creating a spectacle during rainy periods as the water forms dripping curtains around the visitors.

The Pérez Art Museum Miami by Herzog&de Meuron was recently presented with a similar challenge. A hot and humid environment is not typically associated with the calm reflective experience of viewing contemporary art, so the architects had to be creative. By defining a large plinth which spreads out beyond the limits of the internal spaces a greater level of flexibility could be achieved: this plinth is raised up above the limits of potential flooding, creating an almost Miesian podium, and it is doubled high above by a roof structure which creates a shaded micro-climate around the gallery spaces. The somewhat Spartan formal arrangement of the galleries is slowly being transformed by the growth of vines and other plants under the roof, which will not only further improve the atmospheric conditions but will eventually give the museum the appearance of some kind of tropical rain.

It's clear that these new cultural buildings are involved in various strategies for experimenting with forms of artificial landscape, and attempt to firmly stitch these new buildings into their local contexts, even when they are still spectacular object buildings. We can also see investigations of climatic strategies which have been notably absent from the large cultural buildings of the fifteen years, and which will only become more important. But the naturalistic metaphor has been an important part of signature architecture over this period, with many buildings being sold as interpretations of stones, pearls, rivers and other phenomena. Two more recent cultural projects both deploy this kind of thinking, but in a more controlled, perhaps less all-consuming manner than before. The Museum of the History of Polish Jews by Lahdelma&Mahlamäki Architects inevitably bears comparison to the Jewish Museum Berlin of 1996 by Daniel Libeskind. Both are buildings which attempt, through didactic exhibitions, to tell a more complete story of the Jewish who once made up a sizable minority in central and Eastern Europe, before their mass murder in the Second World War. But while Libeskind's building, easily his finest work, offers itself as a hyper-symbolic conceptual object, into which functions and exhibitions had to be fitted, the Polish building concentrates its architectural experimentation into a few very specific features. Located on the site where the Warsaw Ghetto Uprising began, the building is simple in plan, with a glass curtain wall inscribed with Latin and Hebrew texts surrounding it. But running through

部，并为办公室和其他后勤部门提供了充足的常规空间。像华沙历史博物馆一样，峡谷般的空间贯穿整个楼体，创造出一个弯曲折叠的入口立面和不同寻常的循环路线。基于这种有机的形式，建造了照明集成于内的楼梯、屋顶灯和走廊。利用盖里在毕尔巴鄂的古根海姆博物馆项目的设计中使用而闻名的CATIA软件，石板的切割方案和令其就位的钢结构得以由计算机辅助模型事先做好，演示着这些技术的进一步发展，细节的提高和戏剧性的空间效应也变得更为经济。不足为奇，建筑师们讨论地质过程时会寻求引发设计形式的思想。空间的戏剧在此很明显地得到了奖励。我们早已熟悉了文化建筑的细节优化，但在这里，又一次看到了进一步提升，它更关注背景和与周围建筑的关系，同时还试图达到21世纪的建筑需求的刺激效果。

所以，这些建筑代表了21世纪文化建筑的整体方法的改变吗？有这种可能。但很明显，最近的一些建筑在城市入侵的特点上有所节制，更多地考虑了市区和城市条件。这也许意味着有"酋长国风格"的形式和空间的创新与现实环境之间呈现出一种结合。另一方面，我们只能期待在未来的几年里更加注重地标性建筑的气候设计会变得更加重要。大胆的建筑和与自然现象之间的智能交互的进一步妥协，会是一个深受欢迎的发展趋势，但无论目前看到的是最后一个大型工程，或是把这种建筑趋势当作接下来的几年中需要深入的标志或品牌，都为时过早。

the whole building is a huge canyon-like space, created with polished spray-concrete, which resembles rocks carved out by desert winds or perhaps a dried river bed. The concept behind this space is again textual, conceived as the Hebrew phrase "Yam Suf", the parting of the Red Sea, which can signify both the absence of the murdered Jews but also the continuity of the survivors and their culture. But instead of taking over the whole of the architecture, this canyon is performing a symbolic function, without the function of the whole building suffering as a result.

A similar approach, albeit with a much less traumatic story to tell, is in evidence in the Lousiana State Museum and Sports Hall of Fame by Trahan Architects. The building is a fairly conventional box shape, clad in copper pleats which fold out to bring light into the building, and this provides plenty of conventional space for offices and other back-of-house program. But like the Warsaw Museum, a canyon-like space carves through the building, creating a curved and folded entrance facade, and a remarkable circulation route throughout. The stairs, rooflights and corridors are all formed from this organic form, into which all the lighting has been integrated. Utilizing the CATIA software that Gehry made famous in the original Guggenheim Bilbao, the cutting patterns for the stone panels as well as the steel structures that hold them in place were all fully worked out in the CAD model beforehand, demonstrating the further development of these technologies, as detailing improves and the dramatic spatial effects become more cost-effective. As is usual, the architects talk about geological processes as the idea generates the form, but it's obvious that spatial drama is its own reward here. But again, we see here a greater refinement than we are used to in these cultural buildings, with greater care given to their context, their architectural relation to surrounding buildings, while still attempting to achieve the exciting effects that 21st century architecture demands.

So, do these buildings represent a change in the overall approach to the architecture of cultural buildings in the 21st century? Perhaps. It's clear that there is a temperance to the urban aggression of some recent architecture, as the city and its urban conditions are given greater consideration than before. Perhaps this means that there are the beginnings of a synthesis between the formal and spatial innovations of the "Emirate style" and the context of the immediate environment. On the other hand, the greater focus on climatic design within these landmark buildings is something that we can only expect to become more prominent over the coming years, and a further reconciliation between bold architecture and intelligent interaction with natural phenomena can only be a welcome development. But whether we are currently seeing the last of the mega-projects, or whether we will see this tendency towards architecture as logo, as branding, further deepen over coming years, is still too early to tell. Douglas Murphy

De Nieuwe Kolk文化建筑群

De Zwarte Hond

2008年，De Zwarte Hond赢得了在阿森市建造一座多功能文化建筑的设计比赛。场地位于这座古老的城市的边缘，通过设计和建筑施工被改造成一座大型建筑群，拥有剧院、影院、图书馆、咖啡厅、餐厅和公寓的功能。

新文化建筑群是一座由荷兰砌砖、钢和玻璃建成的传统式纪念性建筑。因此它与阿森的环境相契合，而阿森市拥有很多的古老大宅、堡垒和教堂。建筑的入口十分宏伟，位于新剧院广场的附近，而广场是一处富于生机的场地，人们在此停留，以在访问该文化大楼后小饮一杯，心情愉悦。

一面古老的石墙和一座大规模的楼梯将广场与建筑的入口直接连接起来。广场设有一座通往大厅/门厅的大门（人们可在此处参观各个场所）。这是一处参观与被参观的地方。

广场和大厅延续了街道、小巷、广场和花园丰富的连续性的传统，并且将文化功能与阿森市的城市肌理连接起来。

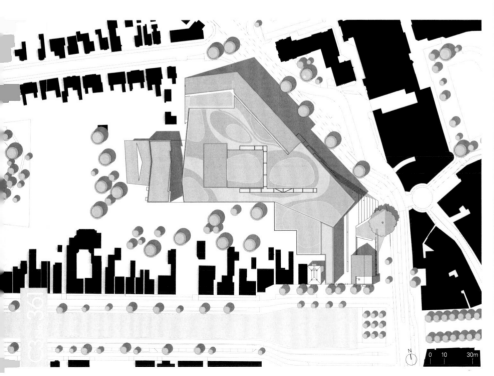

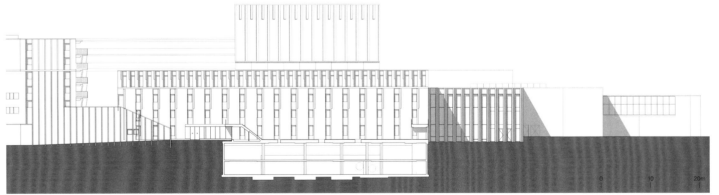

A-A' 剖面图 section A-A'

Culture Complex De Nieuwe Kolk

In 2008, De Zwarte Hond won the design competition for a multifunctional culture building in Assen. The site, at the edge of the historical city has been transformed via a design and building construction into a big complex with functions such as a theater, cinema, library, cafes, restaurants and apartments.

The new cultural complex is a monumental and traditional building made of Dutch brickwork, steel and glass. It therefore fits into the context of Assen which is rich of old mansions, fortresses and churches. The grand entrance of the building is situated next to a new theatre square, which is a lively and pleasant place to stay for a drink after having visited the cultural complex.

A historical stone wall and a staircase of monumental size connect the square directly to the entrance of the building. The square forms a gate to a large lobby/foyer where different venues can be visited. It's the place to see and be seen.

The square and the lobby continue the tradition of a rich sequence of streets, alleys, squares and gardens and connect the cultural program with the urban fabric of Assen.

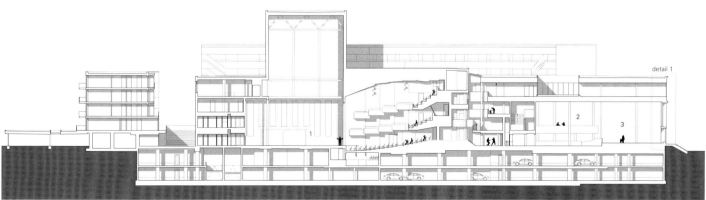

1 大厅 2 图书馆 3 文化建筑入口 1. great hall 2. library 3. culture portal
B-B' 剖面图 section B-B'

1. prefab concrete cover in color
2. prefab concrete ribbon in color
3. aluminium window frame
4. glass panel opaque coated with stainless steel fixation
5. aluminium curtain wall in color
6. shadow box, double glass-inner leaf with color coating
7. aluminium mounting in color
8. aluminium curtain wall in color
9. natural stone plate, fixation by anchors
10. natural stone threshold
11. masonry in tile bond, stabilization with Murfor

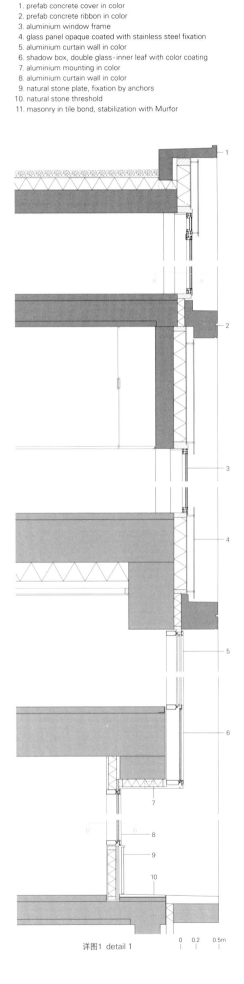

详图1 detail 1

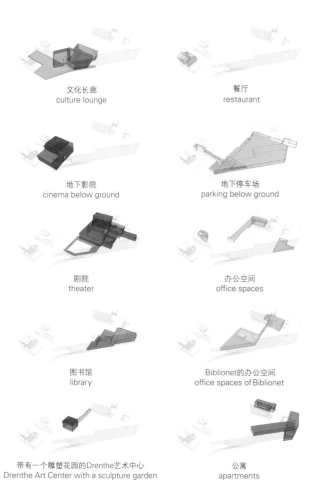

文化长廊 culture lounge	餐厅 restaurant
地下影院 cinema below ground	地下停车场 parking below ground
剧院 theater	办公空间 office spaces
图书馆 library	Biblionet的办公空间 office spaces of Biblionet
带有一个雕塑花园的Drenthe艺术中心 Drenthe Art Center with a sculpture garden	公寓 apartments

项目名称：Culture Complex De Nieuwe Kolk
地点：Weiersstraat 1, 9401 ET Assen, The Netherlands
建筑师：De Zwarte Hond
项目建筑师：Jurjen van der Meer, Tjeerd Jellema
承包商：BAM Utiliteitsbouw
结构工程师：Wassenaar bv
安装：BAM techniek
音效：Peutz
甲方：Gemeente Assen
功能：public library, theater, cinema, center for contemporary art, dwelling, commercial space, cafe, parking garage
总建筑面积：9,243m² / 有效楼层面积：48,000m²
设计时间：2008 / 施工时间：2009 / 竣工时间：2012
摄影师：©Gerard van Beek (courtesy of the architect)

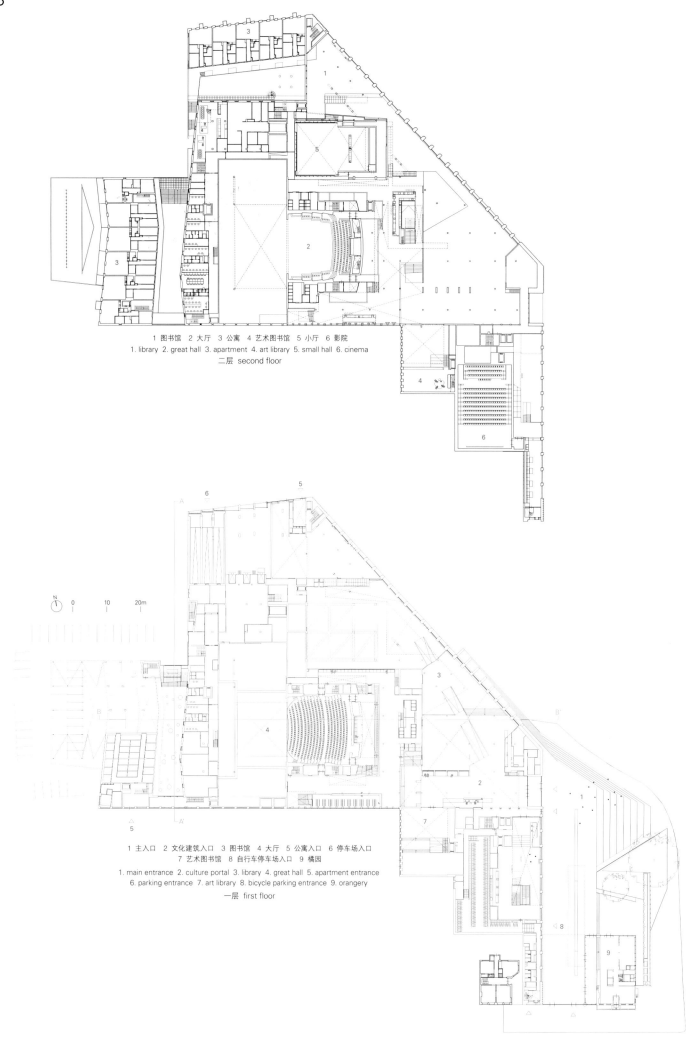

1 图书馆 2 大厅 3 公寓 4 艺术图书馆 5 小厅 6 影院
1. library 2. great hall 3. apartment 4. art library 5. small hall 6. cinema
二层 second floor

1 主入口 2 文化建筑入口 3 图书馆 4 大厅 5 公寓入口 6 停车场入口
7 艺术图书馆 8 自行车停车场入口 9 橘园
1. main entrance 2. culture portal 3. library 4. great hall 5. apartment entrance
6. parking entrance 7. art library 8. bicycle parking entrance 9. orangery
一层 first floor

萨拉戈萨艺术馆

Estudio Carme Pinós

项目启动初期建筑师即面临着两项挑战：第一：设计一座建筑，这座建筑因其独特性及产生的公共空间，可以"感觉像城市一样"。第二：围绕建筑漫步时能看到远处的景观，同时，在其展厅内部还能够提供内省的空间氛围。换句话说，它是一座"像一个城市一样"的建筑，同时能够让身处其间的人们感觉到成为它的一部分。

建筑师通过抬高大厅的方式解决了这两项挑战。这样就释放了地面层的空间，使这里能够容纳更加开放和透明的空间——大厅和商店。建筑师的目标是将其打造成公共空间，让公园穿过建筑下方向城市延伸。夜晚，灯光在此照耀着带有图案的穿孔板外层，另外这处空间还隐藏了支撑被抬高的大厅的结构。

升起的大厅下方是一个半地下的花园，该花园是礼堂的出口，同时也作为前厅和室外餐饮区。这样，由大厅可到达的地下礼堂也可以视为半地下结构，并借由花园与城市直接联系在了一起。

两个悬浮的大厅交错相对，离开一个大厅的时候可以看到另一个大厅下方的城市景观。建筑师认为两个大厅——也就是不同展会之间有必要预留一些减压和放松的空间，为了达到这个目的，建筑师采用了自动扶梯将两个大厅连接起来，为参观者提供了一段可以欣赏到远景的美好旅程——完全不同于以往电梯带来的脱境之感，也不会使参观者感受到任何乘坐电梯所带来的压抑感。

建筑的上部结构是咖啡厅和餐厅，在这里可以观赏城市景观。与之相对的是两个大厅的高度差形成的平台酒吧。酒吧与室内餐厅一起为参观者带来Ranillas河湾和萨拉戈萨世博会园区的美丽景致。

独特而合理的结构使建筑看起来像极了矗立在公园中的一座雕塑作品。

建筑师希望这座建筑可以成为技术进步和文化繁荣的一个标志——只反映我们这个时代最美好的事物。

CaixaForum Zaragoza

The architects start the project by posing two challenges: The first: to design a building that can "feel like a city" – both due to its uniqueness and to the public spaces it generates. The second: to design a building which connects with distant perspectives when walking around, providing at the same time introspection when inside its exhibition halls. In other words, it's a building which "feels like a city" and which makes people feel part of it when they inhabit it.

The architects solve these two challenges by raising the level of the halls. This allows freeing the ground floor, where the architects place the more open and transparent spaces: the lobby and the store. Their aim is to create public spaces, making the park extend into the city by passing under the building – a space which is

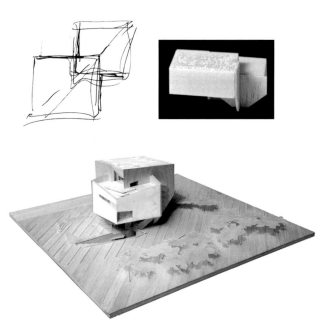

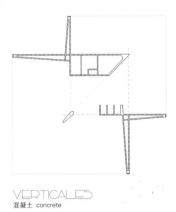
VERTICALES
混凝土 concrete

HORIZONTALES 1
三角形线框 triangulated wireframe

HORIZONTALES 2
■ 混凝土实板 solid slab of concrete
□ 复合板 composite slab

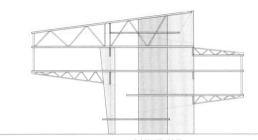
a-a' 剖面图_结构
section a-a'_structure

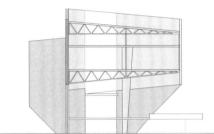
b-b' 剖面图_结构
section b-b'_structure

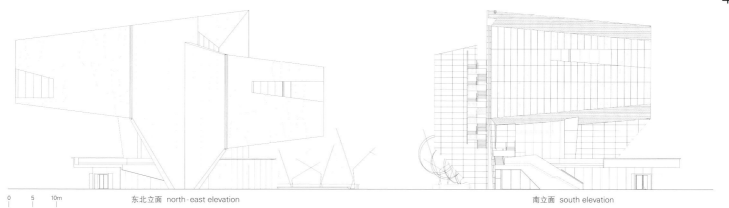

0 5 10m　　　　　东北立面 north-east elevation　　　　　南立面 south elevation

项目名称：CaixaForum Exhibition
地点：Calle José Anselmo Clave, Zaragoza, Spain
建筑师：Carme Pinós Desplat
项目协调员：Samuel Arriola
项目团队：architects_Elsa Martí, Alberto Feijoo, Teresa Lluna, Daniel Cano
结构工程师：Boma
服务工程师：INDUS Ingeniería y Arquitectura SA
音效工程师：Higini Arau
用地面积：9,390m²/总建筑面积：7,062m²/有效楼层面积：1,720m²
设计时间：2008—2010/施工时间：2010—2014
摄影师：
Courtesy of the architect-p.40~41, p.44top, p.46
©Ricardo Santonja (courtesy of the architect) -p.39, p.42, p.43, p.44bottom, p.46, p.47, p.48

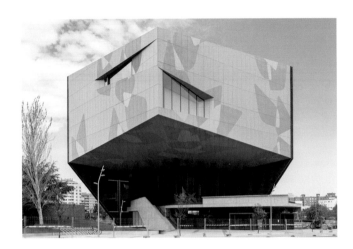

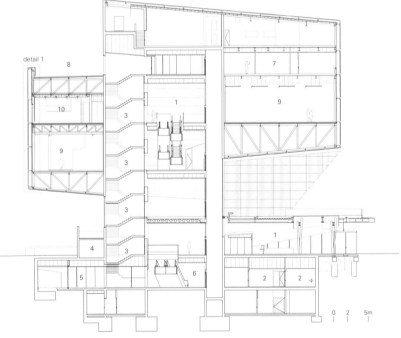

1 大厅 2 员工室 3 主楼梯 4 装卸通道 5 庭院 6 门厅
7 办公室管理/协调区 8 平台 9 展区 10 教室
1. hall 2. staff room 3. main staircase 4. loading and unloading access 5. courtyard 6. foyer
7. office management/coordination 8. terrace 9. exhibition 10. classroom
A-A' 剖面图 section A-A'

1. 3mm aluminium perforated sheet tray
2. tray attachment
3. secondary structure steel pipe
4. structural gusset plate
5. 60mm sandwich panel
6. 50mm rock wool insulation
7. secondary structure galvanized steel facade
8. double glazed unit. curtain wall
9. curtain wall profile in lacquered aluminium
10. curtain wall profile in lacquered steel
11. angle fixing profile
12. standardized profile
13. secondary structure fixing
14. continuous plasterboard suspended ceiling
15. 3mm galvanized steel sheet
16. 3mm lacquered aluminium sheet topping
17. plasterboard profile
18. plasterboard
19. mdf painted skirting board
20. electric skirting board
21. multi-layer resin for external use
22. galvanized steel grating
23. multi-layer resin floor finish
24. 12cm composite slab
25. waterproof coating + geotextile sheet
26. reinforced concrete floor slab
27. flexible polypropylene membrane
28. 50mm rockwool

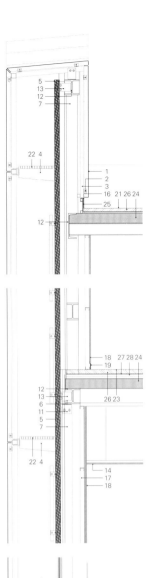

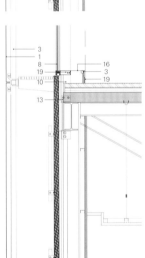

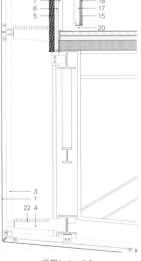

详图1 detail 1

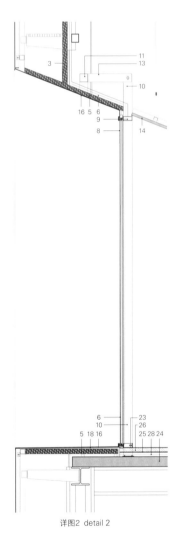

详图2 detail 2

lit at night with drawings obtained by perforating the plate, which in addition hides the structure supporting the elevated halls.

Below the raised halls the architects place a semi-underground garden that serves as the exit to the auditorium and which can also be considered an anteroom or outdoors catering area. Thus, the auditorium – located underground and accessible through the lobby – can be considered as halfway underground and directly connected to the city thanks to the garden.

The two suspended halls face each other at different levels in a way that the visitor who exits one hall has a view of the city below the other hall. The architects believe decompression and relaxation areas are necessary between both halls – i.e., between exhibitions. With this aim, the two halls are connected by escalators, offering a journey which allows visitors to enjoy the distant views – completely different from the decontextualization produced by elevators and which cannot offer the visitor any type of decompression.

On the upper part of the building and with views to the city are the coffee shop and the restaurant. Opposite to them and created by the different levels between halls, there is a terrace-bar which – keeping with the indoors restaurant – allows to enjoy fantastic views of the Ranillas Meander and Expo Zaragoza.

Thanks to its unique and feasible structure, the building appears as a sculptural element amidst the park.

The architects want the building to become a symbol of the progress of the technique and generosity of culture – they want it to be a reflection of only the best things of our times.

Estudio Carme Pinós

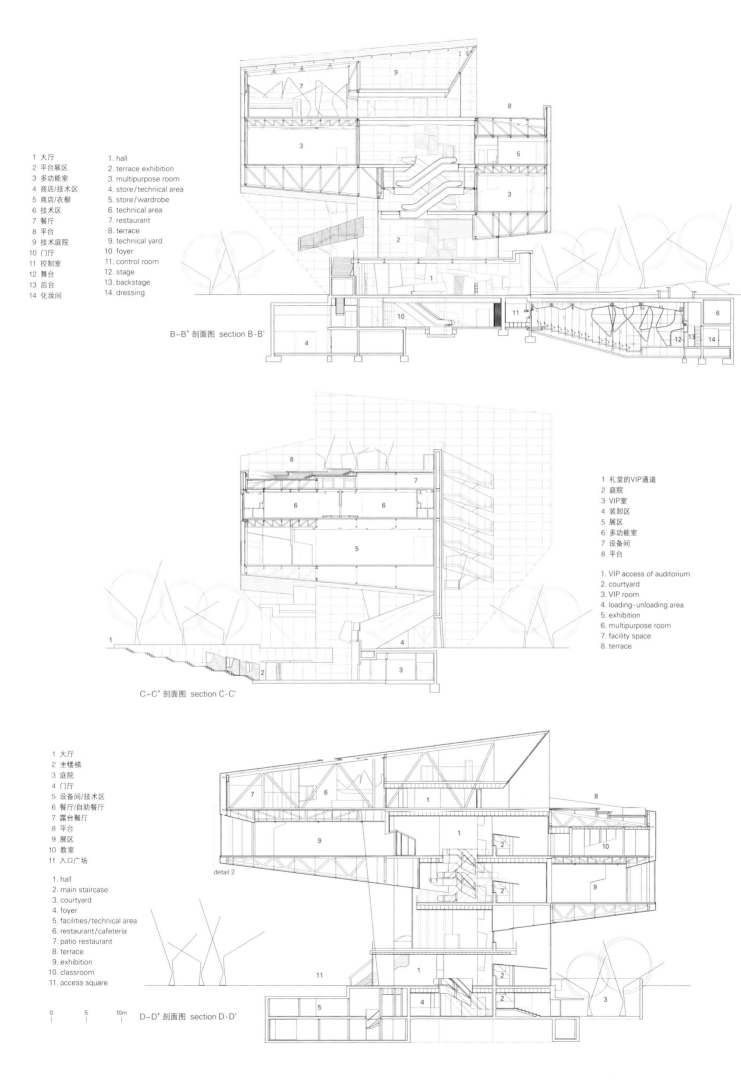

1 大厅	14 餐饮区
2 信息台/衣橱/存物柜	15 VIP室
3 商店	16 卫生间
4 员工室	17 设备间和技术区
5 主楼梯	18 办公室管理/协调区
6 装卸通道	19 餐厅/自助餐厅
7 消防楼梯	20 露台餐厅
8 入口前院	21 厨房仓库
9 入口广场	22 平台
10 庭院	23 大厅的平衡区
11 门厅	24 展区
12 观众区	25 儿童艺术馆
13 控制室	26 教室

1. hall	14. catering
2. information/wardrobe/locker	15. VIP room
3. store	16. toilets
4. staff room	17. facilities / technical area
5. main staircase	18. office management/coordination
6. loading and unloading access	19. restaurant/cafeteria
7. emergency stair	20. patio restaurant
8. access forecourt	21. kitchen warehouse
9. access square	22. terrace
10. courtyard	23. hall counterbalance
11. foyer	24. exhibition
12. audience	25. caixaforum kids
13. control room	26. classroom

五层 fifth floor

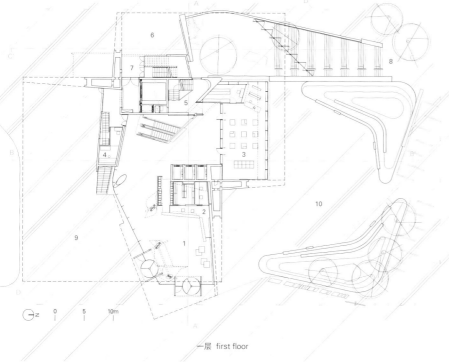

一层 first floor

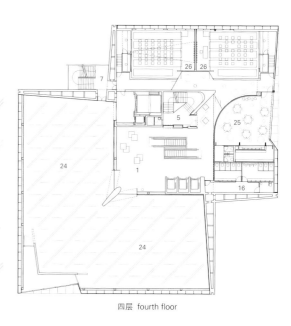

四层 fourth floor

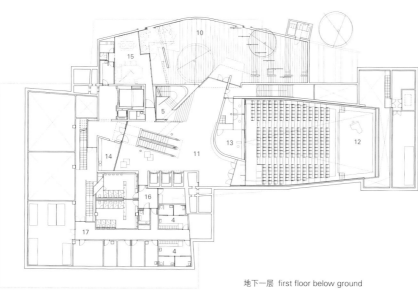

地下一层 first floor below ground

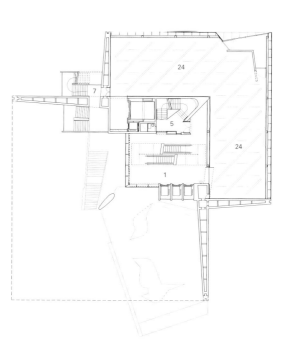

三层 third floor

Eemhuis文化中心

Neutelings Riedijk Architecten

　　Eemhuis文化中心将阿默斯福特市的一系列原有的文化机构结合起来，包括城市图书馆、展览中心、文物档案馆以及一座音乐、舞蹈和视觉艺术学校。该中心位于一处接近市中心的城市再开发区域。项目被规划为一系列的文化功能的堆叠。主要的公共区域全方位地延续到建筑的内部中。在一层，公共广场成为一个覆顶的广场，设有一座大型咖啡馆和通向各个区域的入口。展览中心直接与一层的广场相辉映，且带有一个大型中央展览大厅。大厅位于一层，呈半下沉状态，被一系列纵向排列的小展览室所包围。图书馆是一个阶梯式的信息平台所围成的广场，作为将游客引至主图书馆楼层的城市广场的延续。在楼梯的上方，图书馆"渗入"到一处大型的开放空间，这处空间内放置了书架，设有阅读区和学习区，且可俯瞰整座城市。其上是档案室，且形成这处空间的天花板。建筑的阁楼内设有艺术学校。三个艺术部（戏剧&舞蹈、视觉艺术和音乐）都分别呈现为环绕这座建筑群的悬臂梁形式。立面经过整体规划，成为经典的三部分。柱基由30cm长的加长瓷砖构成，突出了水平衬砌。建筑的顶部由金属嵌板构成，带有星罗棋布的半球形图案，这种图案突出了悬臂体量的间离效果，与荷兰北部的云彩形成对比。

项目名称：The Eemhuis Cultural Center
地点：Eemplein, Amersfoort, The Netherlands
建筑师：Neutelings Riedijk Architecten
结构工程师：ABT Adviesbureau voor Bouwtechniek bv.
建筑物理：DGMR Raadgevend Ingenieurs
安装设计：HE Adviseurs
成本管理：BBN Adviseurs
总承包商：Dura Vermeer Bouw bv.
场地管理和监督：Royal Haskoning DHV
甲方：City of Amersfoort
表面积：16,000m²
设计时间：2006
施工时间：2011
竣工时间：2014
摄影师：©ScagliolaBrakkee (courtesy of the architect)

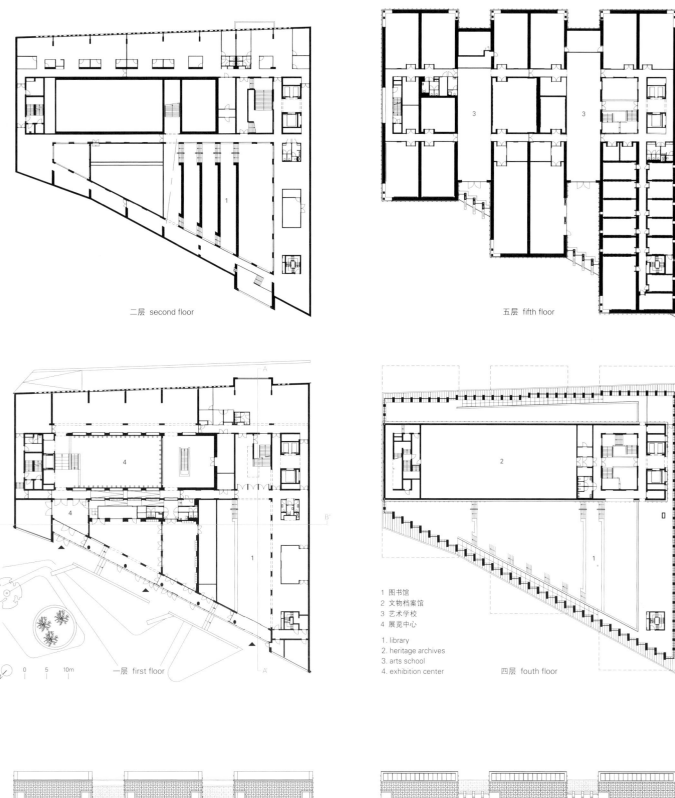

二层 second floor

五层 fifth floor

1 图书馆
2 文物档案馆
3 艺术学校
4 展览中心

1. library
2. heritage archives
3. arts school
4. exhibition center

一层 first floor

四层 fouth floor

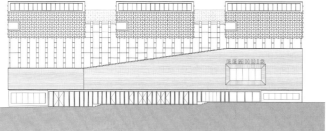

东南立面 south-east elevation

西北立面 north-west elevation

The Eemhuis Cultural Center

The Eemhuis combines a number of existing cultural institutes in the city of Amersfoort: the city library, the exposition center, the heritage archives and a school for dance, music and visual arts. It is located on an urban redevelopment area close to the city center. The building is organized as a stacking of the cultural programs. The public domain is continued into the interior of the building in all directions. At the ground floor, the public square becomes a covered plaza, with a grand cafe and entrances to the various functions. The exposition center is set directly off the square on the ground floor, with a large central exhibition hall that is half sunken in the ground and is surrounded by an enfilade of smaller exhibition rooms. The library is a plaza of stepped information terraces as a prolongation of the city square that brings the visitors up to the main library floor. On the top of the stairs the library spills into a vast open space with book stacks and reading and study areas overlooking the city. Above it hovers the archive volume that forms the ceiling of this space. The attic of the building houses the arts school. The three arts departments (theatre & dance, visual arts and music) are each expressed separately as cantilevered beams that crown the complex.

The facades are composed of a classical tripartite as imposed by the master plan. The plinth is made of 30cm long elongated glazed bricks, reinforcing the horizontal lining. The crown of the building is made of metal panels with a dotted pattern of semi-spheres that enhance the alienating quality of the cantilevered volumes against the northern Dutch clouds.

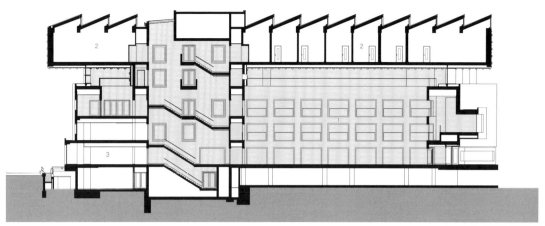

A-A' 剖面图 section A-A'

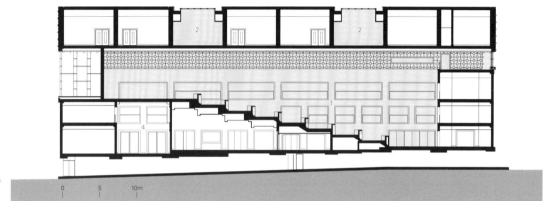

1 图书馆
2 文物档案馆
3 艺术学校
4 展览中心

1. library
2. heritage archives
3. arts school
4. exhibition center

B-B' 剖面图 section B-B'

21世纪国家电影档案馆
Rojkind Arquitectos

位于墨西哥城南部的国家电影档案馆(名为Cineteca)和墨西哥电影学院是拉丁美洲最重要的电影文化遗产所在地。校园占据了Xoco镇内相当大的一部分未被充分使用的区域。这座古镇曾一度被农业用地所环绕,如今又深处城市扩张的中心地带,但由于开发商和市政当局觊觎其优越的地理位置而带来的经济上和政治上的压力,导致其濒临消失。

现存的综合设施在1982年时发生了一场火灾,校园的一部分及档案馆的大部分被烧毁了之后,该建筑作为一座"临时"设施就再也没有被用作合适的用途。除此之外,每天有成千上万的人步行经过这里,往返于城区附近的科约阿坎地铁站。

Cineteca档案馆面临着整体的翻新改建,其最初的项目书包括原有设施的翻修以及额外的保管库和四个放映厅(与现有建筑融为一体)的建造。但是,为了应对迅速发展的城市条件,进一步要做的翻新工作是将该场所的一部分改造成为公共空间,释放Xoco镇新发展环境下充斥着的紧张氛围,缓解行人和游客流量。

首先,地面停车场堆叠成为一座六层高的建筑,这样就释放了这里40%的空间。然后,位于街道对面古镇的公墓处的、方便的"后门入口"被再次利用起来,70%的老顾客会乘坐公共交通工具或步行通过。如今,在翻修后的区域,新项目沿两条中轴线布局,一条垂直于雷亚尔·梅约拉格街道,成为主要的步行入口,另外一条则垂直于墨西哥-科约阿坎街道,供车辆及行人进出。

轴线交叉处成为了一个面积为80m×40m的新公共休闲广场,上面悬挂一个天篷连接着原有的综合设施与新放映厅,为人们遮风挡雨。复合铝板的天篷带有大小不一的三角形穿孔,这样的屋顶结构环绕着新放映室,成为其外立面。天篷下的区域可以作为新、老放映厅间的休息大厅,并设有一些附加的功能,例如:音乐会、戏剧、展览等。

设计在原有项目的基础上增加了一个室外的露天剧场、大片的景观区和新商业区,扩大了社会和文化的互动和交流,给身处这座设施之中的人们一种置身大学校园的感觉。

新放映厅每间有180个座位,老放映厅也借助于现代技术进行了更新。整座设施的室内剧院共提供2495个座位。室外露天剧场可以容纳750人。场地另外增加了两个新的胶片保管库,令Cineteca档案馆的胶片储藏能力又增加了50 000卷。停车量提高了25%,即共有528个车位。

每天成千上万来往于这里的人们如今获得了好客、无束缚的公共空间。附带的便民设施将校园变成了一个受欢迎的聚会地点,让不仅来看电影的人,甚至包括在Xoco镇居住和工作的人们都能够享受置身其中的乐趣。

21C National Film Archive

Located in the southern quadrant of Mexico City, the National Film Archive and Film Institute of Mexico is home to the most important film heritage of Latin America. Its campus occupied an underutilized site of considerable dimensions within the town of Xoco. This historic town, once surrounded by agricultural land,

项目名称：Cineteca Nacional Siglo XXI
地点：Mexico City
建筑师：Rojkind Arquitectos
创始合伙人：Michel Rojkind / 合作者：Gerardo Salinas
项目团队：Gerardo Villanueva, Barbara Trujillo, Alfredo Hernandez, Diego Leal, Andrea Leon, Rodrigo Medina, Philipp Scheuduch, Beatriz Zavala, Birgit Hammer, Juan Manuel Ortuño, David Stalin, Alonso de la Fuente, Rafael Cedillo, / Arie Willem de Jongh, Victor Martínez, Adrian Aguilar, David Guardado. / Media _ Monique Rojkind, Rosalba Rojas, Cynthia Cardenas, Dolores Robles
室内设计：Principal _ Alberto Villareal Bello / Collaboration _ Esrawe Studio / Team _ Isaac Smeke, Felipe Castañeda, Emilia Franssen, Alejandra Hernandez
结构工程师：CTC Ingenieros
屋顶结构工程师：Studio NYL / MEP: IPDS
景观建筑师：Ambiente Arquitectos
音像顾问：Auerbach Pollock Friedlander
音响顾问：Seamonk
照明顾问：Ideas y Proyectos en Luz
平面设计：Citrico + Welcome Branding
施工面积：49,000m²
设计时间：2011 / 竣工时间：2014
摄影师：
©Jaime Navarro(courtesy of the architect) p.70~71, p.71 left
©Paul Rivera(courtesy of the architect) p.62~63, p.65, p.67, p.68~69, p.71 right

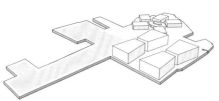

1 原始条件
国家电影档案馆原来有六个放映厅、五个档案保管库以及一个地面停车场,停车场占据了场地42%的面积。

1. original condition
The National Film Archive originally had six screening rooms, five archive vaults and a surface parking that occupied 42% of the total site area.

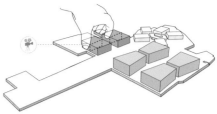

2 放映厅
项目新建了四个全新的放映厅,每个都可供180名参观者使用,总共可为720名观众提供服务。六个原始的放映厅全部进行了翻新和重建,以容纳1775名观众。

2. screening rooms
Four new screening rooms were added and each one has a capacity for 180 viewers for a total capacity of 720. The six original screening rooms were fully updated and renovated with a capacity for 1,775 viewers.

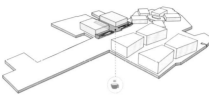

3 夹层
夹层的下面设有一处商业区,作为新放映厅的前厅。

3. mezzanine
Below the mezzanine, a commercial zone is created which also acts as a vestibule to the new screening rooms.

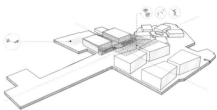

4 屋顶外围护结构/室外露天剧场
新建的放映厅与带有大型屋顶外围护结构的原始放映厅相连,这个屋顶外围护结构下面是一个公共广场,广场在项目的两条主轴线交汇处建成,一座室外的露台剧场可容纳750名观众。

4. roof enclosure / outdoor amphitheater
The new screening rooms are linked to the original ones with a large roof enclosure that covers a public plaza generated in the intersection of the two main axes of the project. An outdoor amphitheater is created with a capacity for 750 viewers.

5 停车场/保管库/植被区
停车场移至一座六层的建筑中,以释放原停车场的场地。新建的公共空间被改造成一个绿意盎然的公共广场。

5. parking / vaults / green areas
The parking is relocated to a six story building to liberate a former parking lot. The new open space is transformed into a green public plaza.

6 影院博物馆/办公室
国家电影档案馆与重建的办公楼和新建的影院博物馆(Taller建筑事务所建造)相得益彰。

6. cinema museum / offices
The National Film Archive is complemented with the renovation of the existing offices building and the creation of the cinema museum by Taller de Arquitectura.

1. original screening rooms
2. amphitheater
3. bookstore
4. restaurant
5. cafe
6. main plaza
7. garden
8. parking garage
9. vaults
10. library and offices
11. cinema museum
12. digital archive screening room

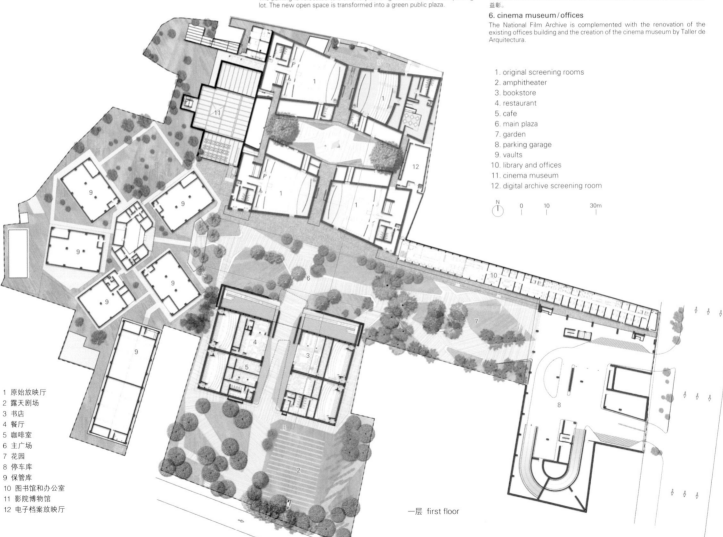

1 原始放映厅
2 露天剧场
3 书店
4 餐厅
5 咖啡室
6 主广场
7 花园
8 停车库
9 保管库
10 图书馆和办公室
11 影院博物馆
12 电子档案放映厅

一层 first floor

now sits deep within the urban sprawl and faces extinction due to economic and political pressures from developers and municipal authorities which covet its privileged location.

The existing complex dated from 1982, when a fire destroyed part of the campus and most of its archive, and was a "temporary" facility never well suited for its purpose. Additionally, thousands of people cross the grounds daily as they walked to and from one of the city's nearby metro station, Estación Metro Coyoacán.

Facing total renewal, Cineteca's original project brief included the expansion and renovation of the existing complex incorporating additional vault space and four screening rooms. But in response to the immediate urban condition, additional restorative work needed to be done to reclaim part of the site as public space, give relief to the dense new-development-filled surroundings of Xoco

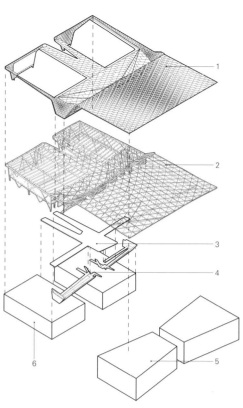

分解轴测图 exploded axonometric

1. roof cladding
2. roof structure
   The roof is made of a steel structure that breaches the gap of the original screening rooms with the new ones.
3. mezzanine
   The new screening rooms are accessed on the mezzanine level.
4. ramps
   The ramps serve as a structural element and usher visitors to the mezzanine level.
5. original screening rooms
   The six original screening rooms were fully updated and renovated.
6. new screening rooms
   Four new screening rooms were added with a total capacity of 720 viewers.

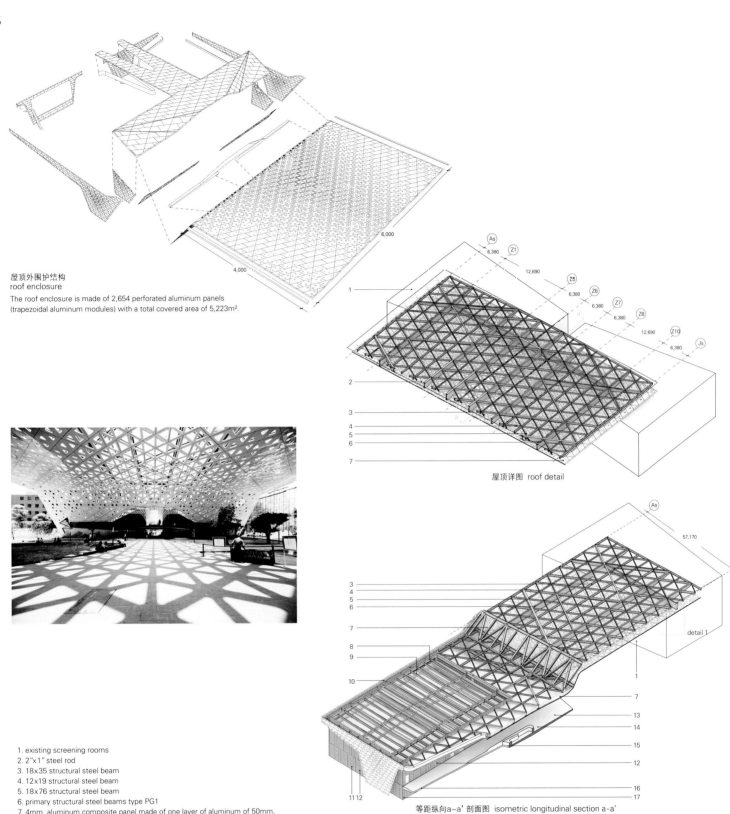

屋顶外围护结构
roof enclosure
The roof enclosure is made of 2,654 perforated aluminum panels (trapezoidal aluminum modules) with a total covered area of 5,223m².

屋顶详图 roof detail

等距纵向a-a'剖面图 isometric longitudinal section a-a'

1. existing screening rooms
2. 2"x1" steel rod
3. 18x35 structural steel beam
4. 12x19 structural steel beam
5. 18x76 structural steel beam
6. primary structural steel beams type PG1
7. 4mm, aluminum composite panel made of one layer of aluminum of 50mm, one layer of polyester with polyethylene and one layer of gloss white aluminum finish
8. primary structure made of TM-15 and TM-17 steel beams with composite metal deck
9. secondary structure made of TM-9 and TM-a beams
10. 12cm, composite deck with brick compression layer and 4.5mm, cement water proofing, finished with gravel and 2" thick insulation
11. new screening rooms
12. black color pre-cast panel mounted over cmu block wall with aluminum anchors
13. metal deck with 5cm, concrete topping 250kg/cm² and 66/1010 wire mesh
14. volcanic slate floor with closed porus type "b" DIM, 30cm x 30cm x 2cm thick with mitered corners @45°
15. 5/16" rolled steel plate counter with tubular 1"x2"x4" steel subframe 12mm, corian base finish color white "glacier" with rounded bull nose
16. perimeter 1/4" steel plate finished with grey "palladio" automotive paint R360
17. 9mm, tempered glass storefront with polished edges
18. clear 6mm, laminated tempered glass with fog pub film + 6mm, tempered clear glass
19. anodized aluminum 2"x3/16" plate angle
20. exterior neutral sealant
21. 4mm x 30cm, aluminum collar

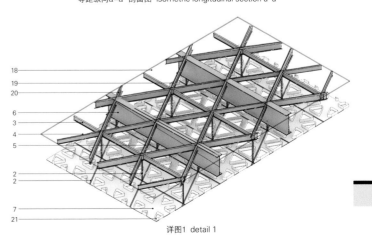

详图1 detail 1

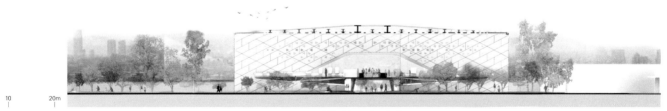

A-A' 剖面图 section A-A'

B-B' 剖面图 section B-B'

and accommodate the constant flow of pedestrians and casual visitors.

First, surface parking was consolidated into a six story structure freeing 40% of the site. Then the pedestrian friendly "back entrance", located across the street from the historic town's cemetery, was reactivated – 70% of Cineteca patrons use public transportation and arrive by foot. The reclaimed space now houses the new programs organized along two axes, one perpendicular to the street of Real Mayorazgo becoming the main pedestrian entrance and the other perpendicular to Av. México-Coyoacán for both car and pedestrian access.

The axes intersection became a new 80m x 40m public plaza sheltered from the weather by a hovering canopy connecting the existing complex with the new screening rooms. Clad in composite aluminum panels, with varied-size triangular perforations, the roof structure wraps around the new screening rooms and becomes their facade. The sheltered space functions as the foyer for the old and new screening rooms and can accommodate additional program options such as concerts, theater, exhibitions, etc..

An outdoor amphitheater, extensive landscaping and new retail spaces were added to the original program expanding the possibilities for social and cultural interaction and exchanges, and giving the complex a university campus feel.

The new screening rooms seat 180 each and the existing screening rooms were updated with current technology. Overall the complex can now seat 2,495 visitors in indoor theaters. The outdoor amphitheater has a 750-person capacity. Two new film vaults were also added to the site, increasing Cineteca's archive capacity by 50,000 reels of film. Parking capacity was also increased by 25% to a total of 528 cars.

The thousands of people that use the grounds everyday now find welcoming unrestricted public space. The added amenities have turned the campus into a favorite gathering space not only for moviegoers but also for Xoco residents and workers.

1. 4mm, aluminum composite panel made of one layer of aluminum of 50mm, one layer of polyester with polyethylene and one layer of gloss white aluminum finish
2. new screening rooms
3. black color pre-cast panel mounted over CMU block wall with aluminum anchors
4. 5/16" rolled steel plate counter
5. 9mm, tempered glass storefront with polished edges
6. 16 gauge steel handrail with primer and automative paint grey "palladio" R360 with 1/2" tube rails mounted @45°
7. polished volcanic slate floor with closed porus type "b" dim. 30 x 30 x 2cm with mitered corners 45°
8. perimeter 1/4" steel plate finished with grey "palladio" automotive paint R-360
9. metal deck with 5cm, concrete topping 250kg/cm² and 66/1010 wire mesh
10. exposed concrete column strength 250kg/cm²

with tubular 1"x2"x4" steel subframe 12mm, corian base finish color white "glacier" with rounded bull nose

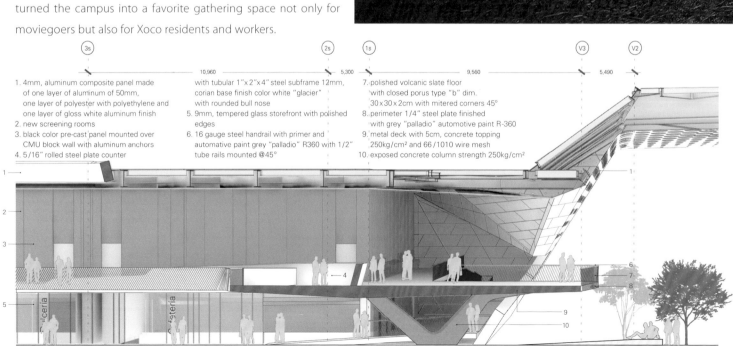

详图2 detail 2

大东文化艺术中心

MAYU architects+ + de Architekten Cie

大东文化艺术中心位于台湾高雄,高雄是台湾南部的经济中心。在中国对外开放市场之前,台湾是中国的主要经济体。城镇进行了发展,成为普遍的工业城市肌理,但却缺乏充足的公共空间。当今,强势品牌"台湾制造"正处在恢复往日荣耀的进程中。这座城市正迫切需要舒适的都市空间来帮助恢复中心城区的活力。因此,大东文化艺术中心的初级构思是一处新公共区域,为城市的舒适等级制定新的高标准。

新文化艺术中心在开放和建筑密集的都市之间建立了强烈的联系,连接凤山河,那里是很受欢迎的公园,同时也是凤山市的历史中心。大东文化艺术中心的设计理念与传统的台湾四合院相似,露台成为居住于此的不同家庭的社会交际中心。四个混凝土体量围成了一个公共广场,广场反过来也将这座综合设施分隔开来。提供了遮阴的广场区域上部的薄膜外层形成了11个巨大的漏斗结构,让人联想到发光的热气球,其尾端接近人行道。

综合设施的配置提高了其公共空间的使用率,被频繁地用于举办具有台湾城区特色的各类活动,比如舞蹈、太极和各种室外游戏。外层在保护游客不受当地强降雨和强烈的日晒等不规律的极端天气侵害的条件下,同时也为各类活动的开展创造了极佳的户外条件。大漏斗的狭窄尾端的开口有利于雨水的排放和新鲜空气的循环。当阳光照射的时候,这些开口又变身成为巨大的天然聚光灯,吸引着人们纷纷不由自主地登台表演。来自不同的社会背景、兴趣爱好和年龄的人们,或独自,或成群结伴,在由长椅和具有花岗岩纹理的下沉广场形成的非正式的圆形平台上表演。

大东文化艺术中心是一处多功能艺术空间,旨在打造一处盛产艺术和文化活动的高端艺术发源地,在对古典音乐和流行文化的艺术才能的培养和训练中起到了至关重要的作用。800个座位的大剧院和一个小彩排厅是这个综合设施的中心部分。艺术中心的四座独立建筑因其X形构架的混凝土立面而统一化,然而每座又由于其内部结构在开放程度上的细微差异而彼此区别。

Dadong Arts Center

The Dadong Arts Center is located in Kaohsiung, the economic center of southern Taiwan. Before the opening of China to foreign markets, Taiwan was the leading economy of the region. Towns developed into a generic industrial city fabric without adequate public spaces. Currently, the strong brand of "Made in Taiwan" is in the process of being restored to its former glory. The cities urgently need pleasant urban space to help regenerate inner-city districts. Therefore, the Dadong Arts Center is conceived in the first place as a new public domain, setting new high standards for the level of urban comfort.

The new arts center creates a strong link between the open and the densely-built cities, connecting the Feng-Shan River, a popular park and the historic centre of the Feng-Shan City. The concept of Dadong Arts Center is similar to the traditional Taiwanese courtyard house, where the patio works as a social exchange for

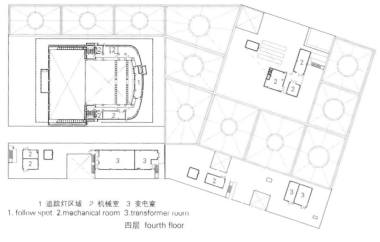

1 追踪灯区域 2 机械室 3 变电室
1. follow spot 2. mechanical room 3. transformer room
四层 fourth floor

1 剧院
2 大型彩排室
3 群组化妆间
4 办公室
5 储藏室
6 商店
7 展览空间
8 室外花园
9 工作室
10 媒体图书馆
11 媒体档案室
12 计算机机房
13 书架区
14 会议室
15 演讲厅
16 艺术工作室
17 VIP室
18 文件储存室
19 控制室

1. theater box
2. large rehearsal room
3. group dressing room
4. office
5. storage
6. shop
7. exhibition space
8. outdoor garden
9. workshop
10. media library
11. media archive
12. computer server room
13. book stacks
14. conference room
15. lecture hall
16. art studio
17. VIP room
18. file storage
19. control room

二层 second floor

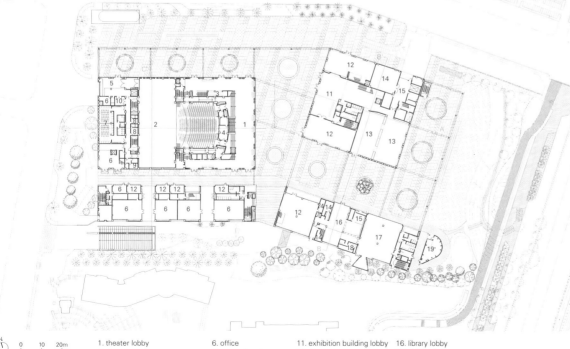

1 剧院大厅
2 舞台
3 售票处
4 照明/声控室
5 装卸区
6 办公室
7 群组化妆间
8 化妆间
9 VIP室
10 钢琴储藏室
11 展馆大厅
12 商店
13 展览空间
14 储藏室
15 工作室
16 图书馆大厅
17 儿童图书馆
18 护理间
19 咖啡室

1. theater lobby
2. stage
3. ticket
4. lighting/sound control room
5. loading area
6. office
7. group dressing room
8. dressing room
9. VIP room
10. piano storage
11. exhibition building lobby
12. shop
13. exhibition space
14. storage
15. workshop
16. library lobby
17. children's library
18. nursing room
19. cafe

一层 first floor

项目名称：Dadong Arts Center/ 地点：Kaohsiung, Taiwan
建筑师：MAYU architects, de Architekten Cie
项目建筑师：MAYU_Malone Chang, Yu-lin Chen / Cie_Branimir Medić, Pero Puljiz
项目团队：MAYU_Kwantak AUYEUNG, Yachih KUO, Fenlan CHEN, Mavis LIU, Yayun WANG, J. Hsiu, J. Yang, W. Lo, Y. Mai, C. Chen, Y. Lee, H. Shen, I. Shr, Y. Huang, R. Huang, B. Guo, S. Wang / Cie _ V. Ulrich, M. Ismael, T. Cheng, Ron Garritsen, Albert van Gelderen, L. Cvetko, H. Gladys, C. Eickelberg, M. A. Rival
施工监督：MAYU_Wei Cheng LI, Yonghao CHEN, Chih-Hung WANG, Wanzhen CHEN, Qi Yang HUANG, Binghong MA
结构工程师：Arup Amsterdam, Tien-Hun Engineering Consultant Inc.
音响顾问：Peutz & Associates, Gade & Mortensen Akustikk, Wei-Hwa Chiang NTUST
环境技术顾问：Hander Engineering & Construction Inc., I.S. Lin & Associates
甲方：Kaohsiung City Government
功能：multipurpose theater, rehearsal hall, exhibition hall, library, administration office, outdoor activity space
用地面积：30,356m² / 建筑面积：8,822m² / 有效楼层面积：36,470m²
竞赛时间：2007.1 / 施工时间：2008.9 / 竣工时间：2012.3

摄影师：
©Jui-Tsung Pan(courtesy of the architect)-p.74 bottom, p.78 top, p.83
©Guei-Shiang Ke(courtesy of the architect)-p.72~73, p.74 top, p.76~77, p.80

©Branimir Medic(courtesy of the architect)

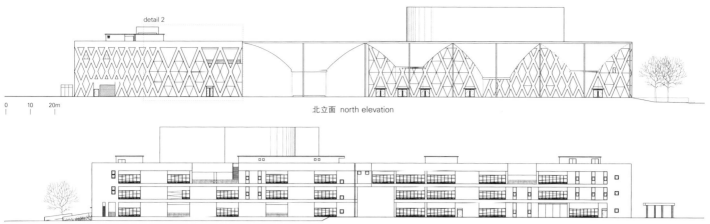

北立面 north elevation

南立面 south elevation

figure 1: membrance stress in warp direction - model 1,
load case 5 - maximum 22.15kN/m

figure 2: membrance stress in warp direction - model 1,
load case 5 - maximum 20.32kN/m

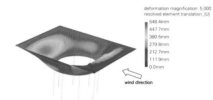

figure 3: deformation of the membrane - model 1,
load case 5 - maximum 548mm

figure 4: forces in the membrane cable - model 1,
load case 5 - maximum 33.44kN

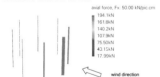

figure 5: forces in the prestressed cables - model 1,
load case 5 - maximum 195kN

figure 6: Mzz in the lower steel ring - model 1,
load case 5 - maximum 104kNm

figure 7: stresses in the lower steel ring - model 1,
load case 5 - maximum 106 N/mm²

薄膜分析图 membrane analysis

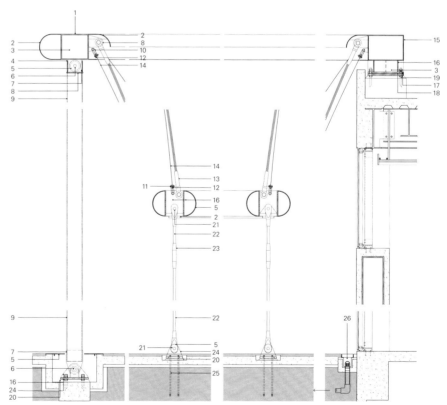

详图1 detail 1

1. PL 12~19t (ASTM A572)
2. cover PL-4.5t
3. RB.PL 12~16t (ASTM A572)
4. ring PL-6t
5. PL-9t
6. stainless steel bolt 50ø (M-12)
7. PL-22t
8. PL-16t
9. P: 273ø×6.35t (ASTM A53 GR.B)
10. U-profile steel @400(3.2t)
11. M12 bolt @200(SUS)
12. 2-50X9t extruded aluminum
13. cable 20~22ø(GALV+coating) SUS 316 end fitting
14. PTEE
15. □:650x450x12t (ASTM A-572)
16. PL-12t
17. PL-20t
18. 2-PL-2t(SUS)
19. 6xM36 embedded bolt(ASTM-A307) (depth 600mm)
20. non-shrinkage cement
21. stainless steel bolt 30ø~42ø (M12)
22. tension rod 28ø~39ø (SUS 316)
23. pipe turnbuckle (SUS316())
24. PL-30t
25. 4-M30 embedded bolt (depth 600mm)
26. drainage

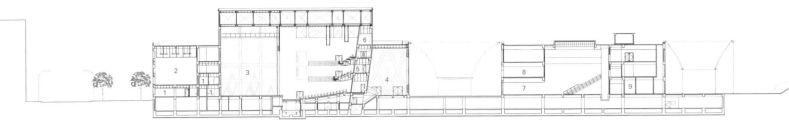

1 化妆间 2 大型彩排室 3 舞台 4 剧院大厅 5 剧院 6 追踪灯区域 7 展馆大厅 8 大厅 9 展区
1. dressing room 2. large rehearsal room 3. stage 4. theater lobby 5. theater box 6. follow spot 7. exhibition lobby 8. lobby 9. exhibition

A-A' 剖面图 section A-A'

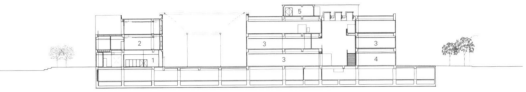

1 图书馆大厅 2 媒体图书馆 3 展区 4 储藏室 5 机械室
1. library lobby 2. media library 3. exhibition 4. storage 5. mechanical room

B-B' 剖面图 section B-B'

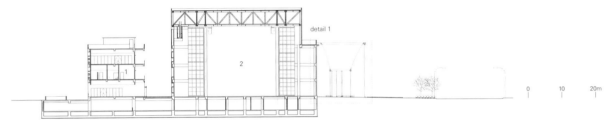

1 储藏室 2 舞台   1. storage 2. stage

C-C' 剖面图 section C-C'

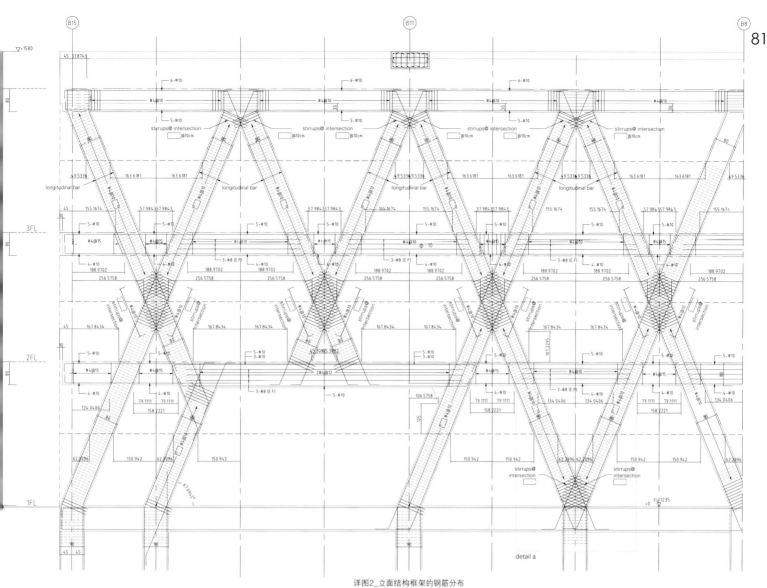

详图2_立面结构框架的钢筋分布
detail 2 _ reinforcement arrangement of facade structural frame

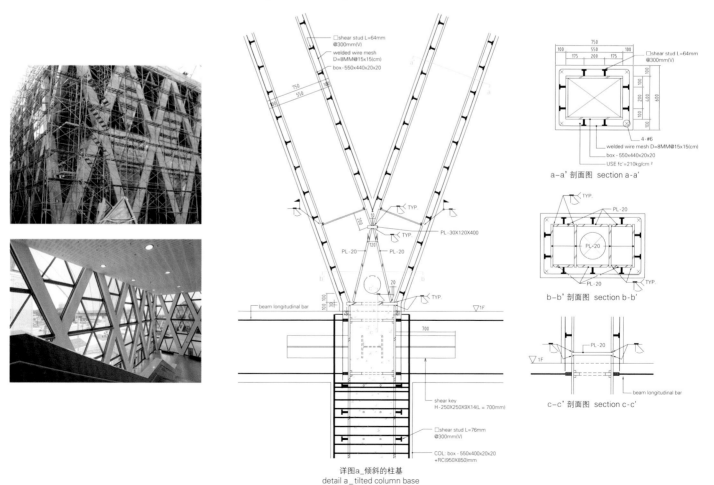

详图a_倾斜的柱基
detail a _ tilted column base

a-a' 剖面图 section a-a'

b-b' 剖面图 section b-b'

c-c' 剖面图 section c-c'

the different families living around it. Four concrete volumes line a public plaza that also penetrates the complex to separate the buildings from one another. Providing shelter, a membraneous skin above the plaza areas features 11 enormous funnels, reminiscent of illuminated hot-air balloons, with tapered ends dipping to face the pavement.

The configuration of the complex stimulates the intense use of public space by activities such as dance, tai chi and various games which characterize outdoor areas in Taiwanese cities. The skin protects visitors from the region's periodically strong rains and intense sunshine, while creating a perfect outdoor situation for such activities. Openings in the narrow ends of the funnels allow for the drainage of rainwater and the circulation of fresh air. When the sun shines, these openings turn into enormous natural spotlights, enticing people to act. All day long, people of different social backgrounds, interests and ages – individually as well as in groups – are performing on the informal circular podia formed by benches and granite-lined basins.

The Dadong Arts Center is a multi-functional art space, aiming to be a state-of-the-art-venue for producing artistic and cultural events. It also plays an important role in cultivating and training artistic talents in classical as well as popular culture. A 800-seat theater and a small rehearsal hall are the center piece of the complex. The X-framed concrete elevations of the four individual buildings of the arts center are homogenous, yet each one indicates the programme inside by subtle changes in the degree of openness of the structure.

Courtesy of the architect

迈阿密佩雷斯艺术博物馆

Herzog & de Meuron

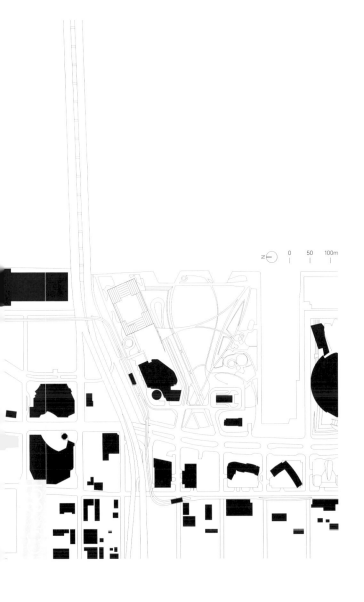

新的迈阿密佩雷斯艺术博物馆（PAMM）位于博物馆公园内，是比斯坎湾市中心海滨重建项目的一部分。它毗邻帕特里夏与菲利普·弗罗斯特科学博物馆，并和一条连接迈阿密陆地和海滩的主要高速公路相邻。新PAMM同时面向公园、滨水和城市，是一座纳四方宾客的开放性结构建筑。

迈阿密以标志性的装饰派艺术区而闻名。然而真正令迈阿密如此不凡的原因是其得天独厚的气候条件、茂盛的植被和多元的文化。如何能够充分利用这些宝贵的地方资源并将其转化到建筑之中呢？

就像本刊早前的文章中提到的一些例子，比如纳帕谷的多米诺斯葡萄酒厂，建筑周边的环境条件本身成为了建筑构思的中心。由于靠近水，博物馆被抬离地面，这样艺术品就处在风暴潮水位之上。建筑师利用建筑下方的空间作为露天停车场，有利于采光及通风，同时还能处理雨水径流。从停车场那一层支起的支柱支撑着博物馆平台，也成为支撑遮阳雨篷的柱子，而雨篷覆盖了整个场地，形成了一处阳台般的公共空

间,欢迎来到博物馆和公园的游客们。面向海湾,一段宽阔的台阶将平台与滨水步道相连。该建筑最突出的特点是其天篷、平台、柱子、植被;也就是说,占据了整个场地的阳台。博物馆的内部体量嵌套其中,悬浮于整个结构框架之中,每层都呈现为其所需的形状。因为美术馆没有任何给定的形式,建筑师必须与博物馆的员工密切合作,自由发挥来开发策展布局,最优化配置的展区除了能够容纳逐渐增加的收藏品外,同时也能够为临时展览提供充足的空间。

PAMM的美术馆设计为四个不同的主题类型:综览美术馆、焦点美术馆、项目美术馆和特展区。它们占据了一层的一部分和整个二层。综览美术馆展示了博物馆的收藏品,作为其他几类美术馆之间的连接组织。建筑师流动地将它们连接成一个非线性序列,使彼此之间形成空间上的联系。这些美术馆的特点是有大大的开口,从那里可以俯瞰到公园、迈阿密市中心、海湾和高速公路。沿着房间流动的顺序,单一的封闭空间不时地被窗口打断,展现出一位艺术家本身、一个主题、一个特殊的收藏品,或一项受委托的任务。这些空间被称为焦点美术馆和项目美术馆。第四类,特展区被设计为宽敞的展厅,以适应当代艺术展览。综览美术馆、焦点美术馆及项目美术馆在建筑内形成了一个严格而有节奏的序列,在与外部的比例和关系上有所差异,而特展区则是灵活的,只有较少的开口通到外面,并且可以通过临时的墙壁进行细分。

在建筑的中心,一段和美术馆一样宽大的楼梯连接着两个展区的楼层。这段楼梯同时还作为礼堂,使用了隔音的幕布来为讲座、电影放映、音乐会和演出这些不同的场合提供空间。建筑师的想法是避免将这些活动独立分设在各个空间中,导致这些空间在大部分时间都闲置不用。在PAMM,人们时常可见为了各项活动而开展的准备工作。在没有活动举办的期间,游客和员工则在空间内阅读,或向参观团体做介绍等等。博物馆的商店和酒馆设在平台那一层,面向海湾。教学研究设备和博物馆办公室均位于三楼。建筑师将这些共享空间设置在建筑的外围,意在使其最大限度地曝露于阳台、比斯坎湾和博物馆公园。

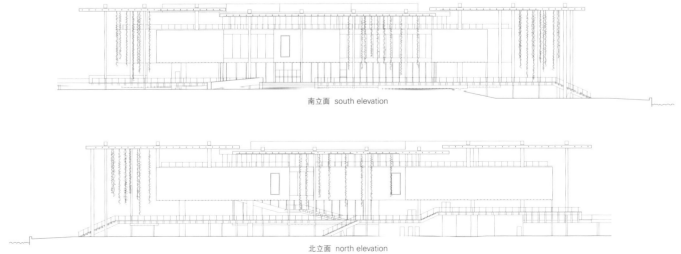

南立面 south elevation

北立面 north elevation

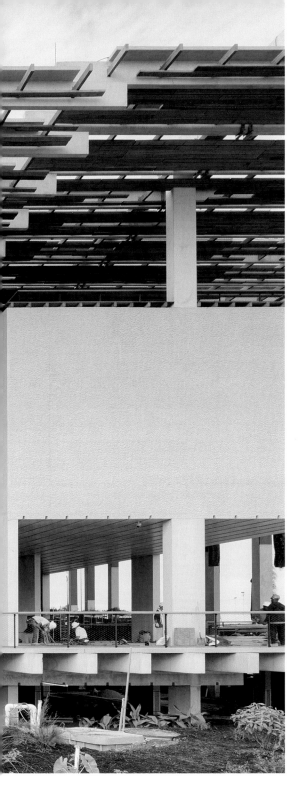

Pérez Art Museum Miami

The new Pérez Art Museum Miami(PAMM) is located in Museum Park, part of the redeveloping downtown waterfront on Biscayne Bay. Its direct neighbors are the Patricia and Phillip Frost Museum of Science and a major freeway, connecting mainland Miami with Miami Beach. Simultaneously oriented towards the park, the water and the city, the new PAMM is an open and inviting structure from all sides alike.

Miami is known for its iconic art deco district. What makes Miami so extraordinary however, is its amazing climate, lush vegetation and cultural diversity. How can these assets be fully exploited and translated into architecture?

As in previous examples of our work, such as the Dominus Winery in Napa Valley, the building's environmental circumstances become central to its architectural concept. Due to its proximity to the water, the museum is lifted off the ground for the art to be placed above storm surge level. The architects use the space underneath the building for open-air parking, exposed to light and fresh air that can also handle storm-water runoff. Rising from the parking level, the stilts supporting the museum platform become columns supporting a shading canopy, which covers the entire site creating a veranda-like public space that welcomes visitors to the museum and the park. Facing the bay, a wide stair connects the platform to the waterfront promenade.

The expression of the building comes from the canopy, the platform, the columns, the vegetation: in other words, the Veranda occupying the entire site. The museum's interior volume nests within it, suspended amid the structural framework, each floor assuming the shape it needs. Because the galleries did not have to fit into any given form, the architects had the freedom to develop the curatorial layout, in close collaboration with the museum staff, to what felt like an optimal configuration to exhibit and develop the growing collection as well as to provide ample space for temporary exhibitions.

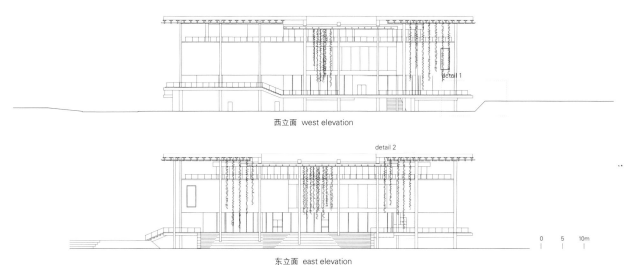

西立面 west elevation

东立面 east elevation

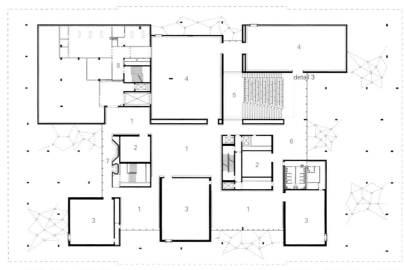

1 综览美术馆 2 项目美术馆 3 焦点美术馆 4 特展区 5 礼堂 6 举办特殊活动的开间 7 游客美术馆 8 后勤区
1. overview gallery 2. project gallery 3. focus gallery 4. special exhibition gallery
5. auditorium 6. special events bay 7. visitors' gallery 8. back of house
二层 second floor

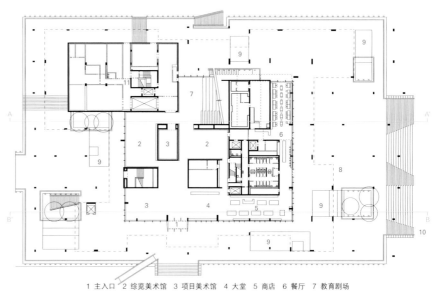

1 主入口 2 综览美术馆 3 项目美术馆 4 大堂 5 商店 6 餐厅 7 教育剧场
8 门廊 9 花槽 10 海湾一侧的楼梯 11 后勤区
1. main entrance 2. overview gallery 3. project gallery 4. lobby 5. store 6. restaurant
7. education theater 8. porch 9. planter 10. bayside stair 11. back of house
一层 first floor

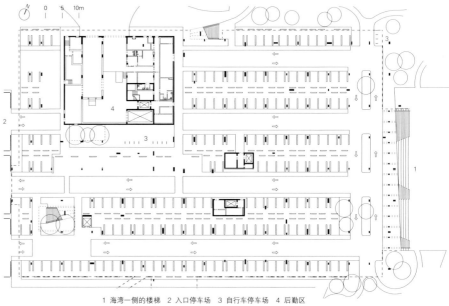

1 海湾一侧的楼梯 2 入口停车场 3 自行车停车场 4 后勤区
1. bayside stair 2. entrance parking 3. bike parking 4. back of house
地下一层 first floor below ground

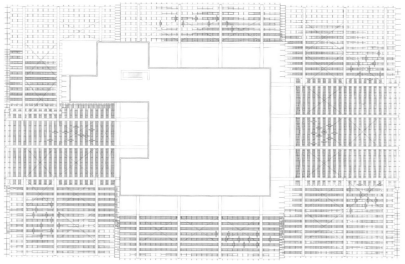

屋顶 roof

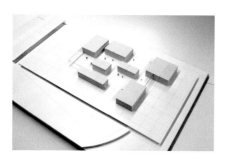

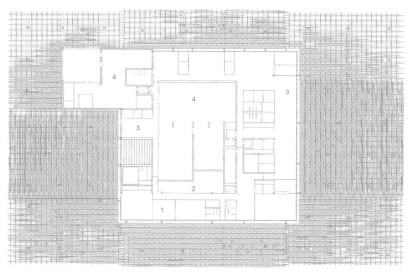

1 培训区 2 图书馆 3 办公区 4 后勤区
1. education area 2. library 3. office area 4. back of house
三层_反射天花板平面 third floor _ reflected ceiling plan

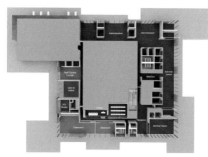

三层 level 3

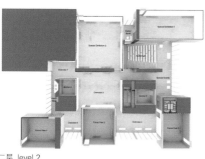

二层 level 2

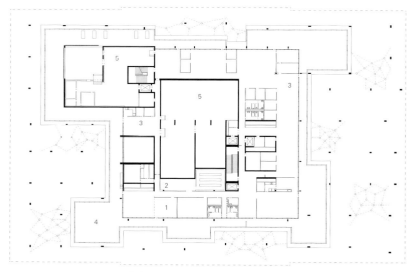

1 培训区 2 图书馆 3 办公区 4 平台 5 后勤区
1. education area 5. library 3. office area 4. terrace 5. back of house
三层 third floor

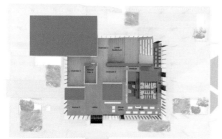

一层 level 1

PAMM is organized around four different gallery types: Overview, Focus, Project and Special Exhibition galleries. They occupy part of the first and the entire second floor. The Overview galleries, displaying the museum's collection, serve as the connecting tissue between the other gallery types. Fluidly connected in a non-linear sequence, they allow relationships to be formed between spaces. They are characterized by large openings with views onto the park, downtown Miami, the bay and the freeway. Along this flowing sequence of rooms, single enclosed space punctuated by windows shows an individual artist, a theme, a specific collection or a commissioned work. These spaces are called Focus and Project galleries. The fourth type, the Special Exhibitions galleries function as spacious exhibition halls designed to accommodate contemporary art exhibitions. The Overview, Focus and Project galleries form a firm and rhythmic sequence through the building, varying in proportion and relationship to the outside. On the other hand, the Special Exhibitions galleries are flexible, with fewer openings to the outside and can be subdivided by temporary walls.

At the heart of the building, a stair as large as a gallery connects the two exhibition levels. This stair also functions as an auditorium, using sound-insulating curtains in different configurations to provide space for lectures, film screenings, concerts and performances. The architects' idea was to avoid for such events to be isolated in a space remaining unused for most of the time. At PAMM, events in preparation are visible. When the space is not actively used for events, it is used by visitors and staff for individual readings, introductions to groups and the like. The museum shop and bistro are located on the platform level and are oriented to the bay. Education and research facilities are on the third floor along with the museum's offices. The architects place these communal spaces at the periphery of the building, maximizing their exposure to the Veranda, Biscayne Bay, and Museum Park.

Herzog & de Meuron

项目名称：Pérez Art Museum Miami
地点：1103 Biscayne Boulevard, Miami, FL 33132, USA
建筑师：Herzog & de Meuron
主要合作者：Jacques Herzog, Pierre de Meuron, Christine Binswanger
项目团队：Associate_Charles Stone
Associate, Project manager_Kentaro Ishida, Stefan Hoerner
Workshop_Adriana Mueller, Ahmad Reza Schricker, Daekyung Jo, Dara Huang, Günter Schwob
Associate_Hugo Moura, Ida Richter Braendstrup, Jack Brough, Jayne Barlow
Workshop_Jason Frantzen, Jeremy Purcell, Joana Anes, Margarida Castro, Masato Takahashi, Mehmet Noyan, Nils Sanderson, Roman Aebi Silja Ebert, Sunkoo Kang, Valentine Ott, Wei Sun, Yuichi Kodai, Yuko Himeno
成本顾问：Stuart-Lynn Company
设计顾问：Herzog & de Meuron
执行建筑师：Handel Architects
空中花园设计师：Patrick Blanc
暖通空调工程师：Arup
景观设计师：GEO Architectonica
管道/消防管理者：JALRW, Doral
结构工程师：Arup
本地结构工程师：Douglas Wood, Coral Gables
功能：exhibition, educational, auditorium, retail, dining and refreshment, back of house, office
用地面积：14,221m²
总建筑面积：8,598m²
有效楼层面积：11,125m²
设计时间：2006.10—2009.9
施工时间：2010.11—2013.10
摄影师：©Iwan Baan (courtesy of the architect)

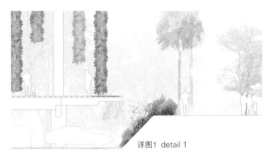

详图1 detail 1

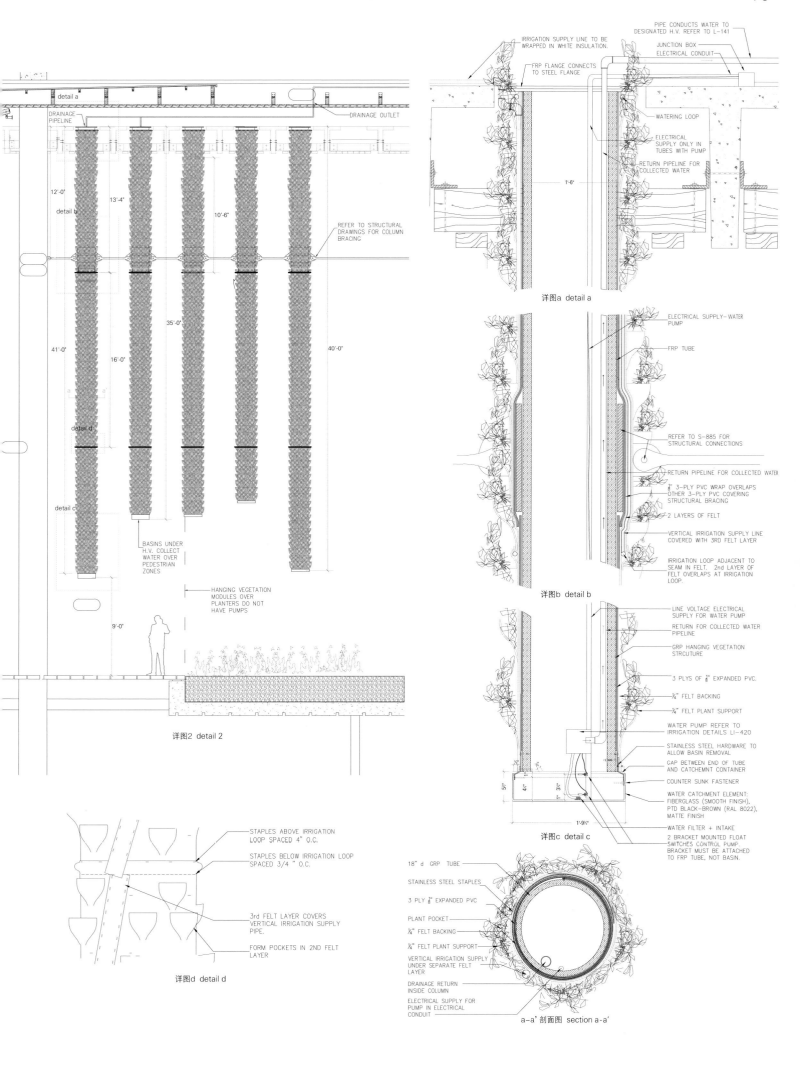

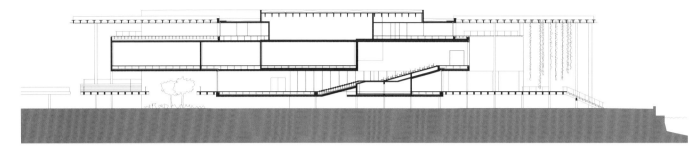

A-A' 剖面图 section A-A'

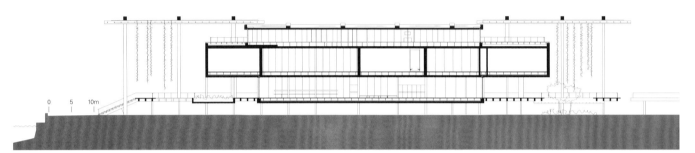

B-B' 剖面图 section B-B'

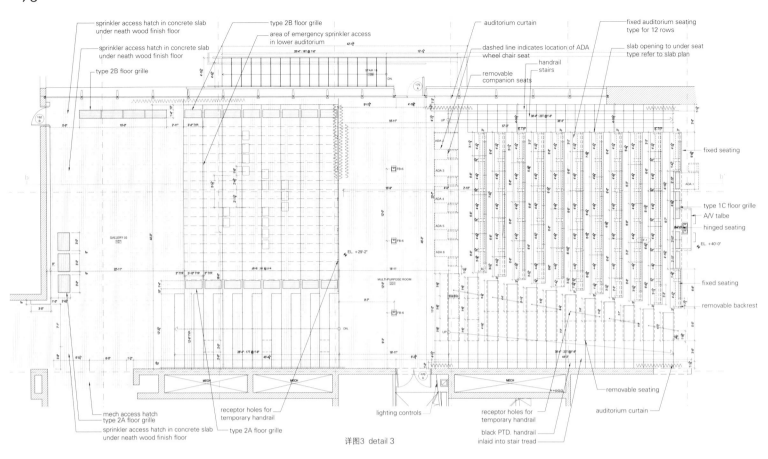

详图3 detail 3

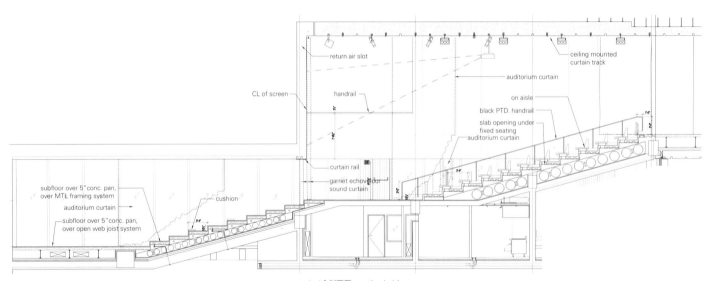

b-b' 剖面图 section b-b'

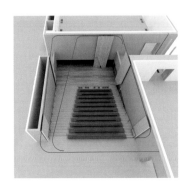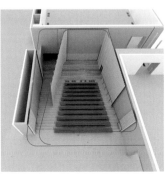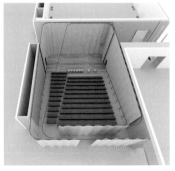

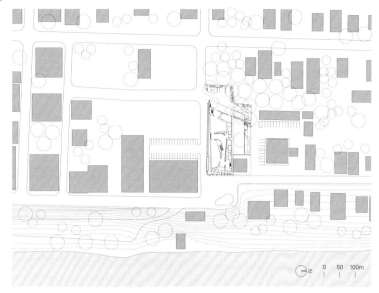
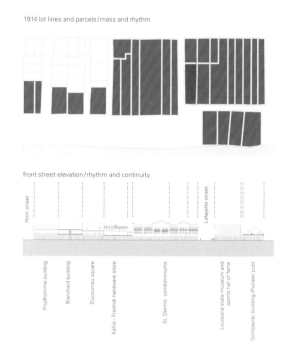

路易斯安那州立博物馆和体育名人堂
Trahan Architects

位于历史悠久的纳什托基的路易斯安那州立博物馆和体育名人堂，融合了之前位于大学体育馆和一个19世纪的法院内的对比强烈的收藏品，提升了访问者对二者的体验。坐落于路易斯安那州历史悠久的凯恩河岸边的一处聚居地（被政府收购），这个设计实现了在历史与现实，过去与未来，包容与被包容之间的对话。

建筑师的探索集中在三个问题上。即设计该如何钻研甲方的项目书，以展示运动和历史？建筑如何与历史建筑的构造相对应？建筑如何与当地的环境相联系？

首先，建筑师的解决办法是将运动阐释为文化历史的一部分，而不是独立的主题。运动和地方历史可能吸引不同的观众，同时展示活动和建筑结构努力探索运动和历史之间的联系。建筑空间在视觉上和结构上自然流动，形成了能够体现最高水平的展览、教育和配套设施的功能。访问者既能独立又能同时体会两座建筑所讲述的东西。

其次，历史风格的作品被留出来表达一种回应场地的设计语言。内部组织结构是蜿蜒的城市流线的延伸，且这个设计将历史商业中心的特征与规模和周边居住环境融合在一起。"简单"的外部覆盖了起褶的铜板，与周围种植园的百叶窗和墙板相呼应，和内部的优雅曲线形成对比和补充。百叶窗的表皮控制着光线、景色和通风，使建筑立面富于生机，并形成了建筑表面的连接，而这以前是通过建筑的装饰物来实现的。人们在入口处就可以看见流线形的内部，吸引着访问者走到内部焕然一新的展示空间。

第三，设计反映了古代河流的鬼斧神工，河床地貌是内部流动形设计的灵感来源。动态的门厅由1100块铸石面板雕刻而成，将所有系统无缝式地融为一体，并且由上部的自然光营造出流光溢彩的效果。白色的石头中间采用一种填充物，材料为17世纪的居民使用的马鬃、泥土和西班牙生长的寄生藤。流动的表面直抵长廊，作为电影和展览的屏幕。沿着小径到达上层的最高点，有一个阳台俯瞰整座城市的广场，上有铜质天窗遮挡，进一步连接着内部和公共空间。

Louisiana State Museum and Sports Hall of Fame

The Louisiana State Museum and Sports Hall of Fame in historic Natchitoches, Louisiana merges two contrasting collections formerly housed in a university coliseum and a nineteenth century courthouse, elevating the visitor experience for both. Set in the oldest settlement in the Louisiana Purchase on the banks of the Cane River Lake, the design mediates the dialogue between sport and history, past and future, container and contained.

The architects' exploration focuses on three questions. How does the design explore the client brief to exhibit sports and history simultaneously? How does it respond to the historic building fabric? How does it make a connection to context?

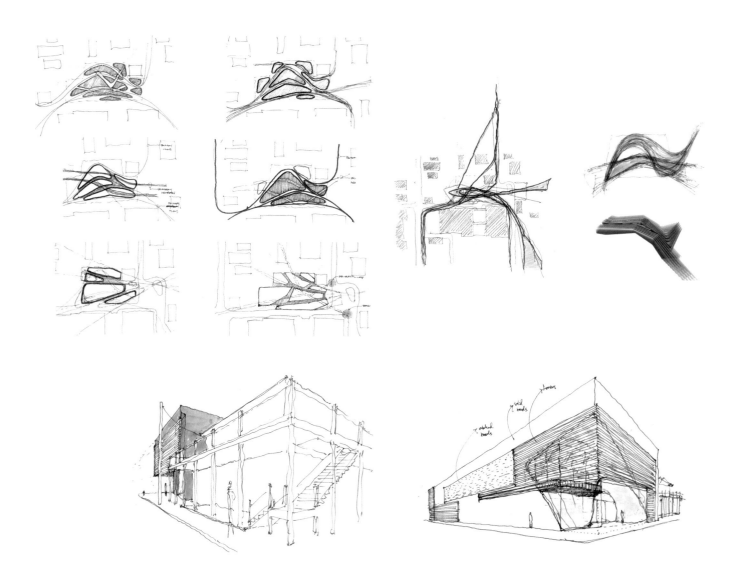

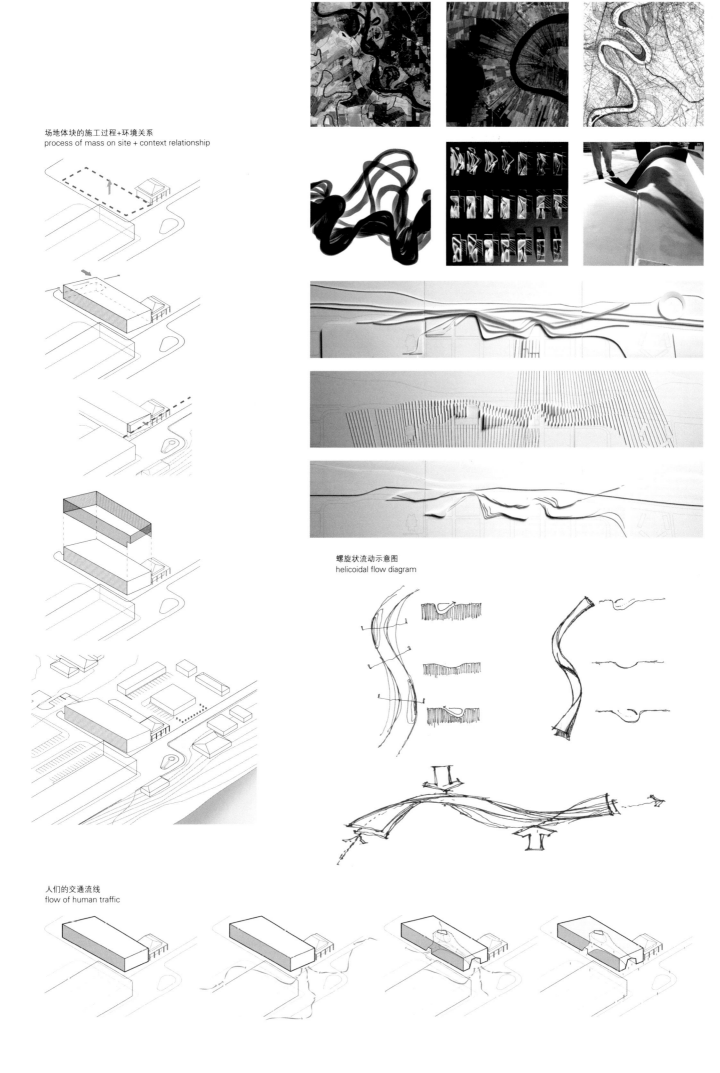

立面和图案研究
elevation and patterning studies

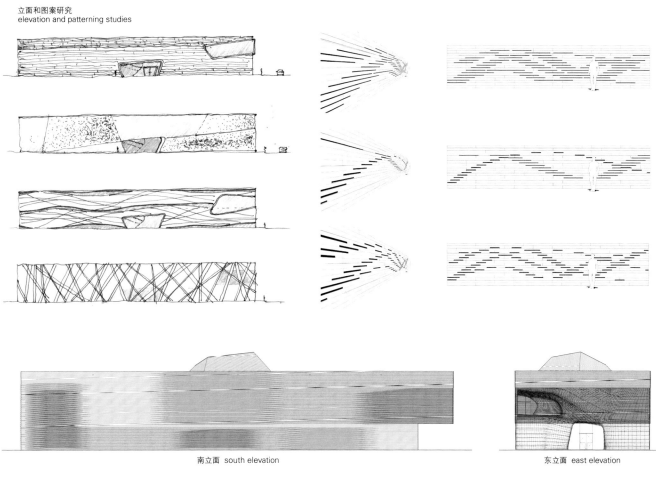

南立面 south elevation

东立面 east elevation

0 10 20m

北立面 north elevation

西立面 west elevation

项目名称：Louisiana State Museum and Sports Hall of Fame
地点：Natchitoches, Louisiana, USA
建筑师：Trahan Architects
设计主管：Victor F.
项目建筑师：Brad McWhirter
设计师：Ed Gaskin, Mark Hash, Michael McCun
项目团队：Sean David, Blake Fisher, Erik Herrmann, David Merlin, Benjamin Rath, Judson Terry
室内设计师：Lauren Bombet Interiors
机械/电气/管道/消防工程师：Associated Design Group
结构工程师：LBYD
土木工程师：CSRS
地热技术工程师：GeoConsultants
总承包商：VCC
景观建筑师：Reed Hilderbrand Associates
建筑信息模型管理和技术指导：Case
铸石和钢支架外形设计和详图设计：Method Design
铸石和钢支架结构工程师：David Kufferman PE
音效工程师：SH Acoustics
防水设计师：Water Management Consultants & Testing, Inc.
用地面积：3,449m²
总建筑面积：2,601m²
有效楼层面积：1,300m²
造价：USD 12,600,000
设计时间：2010
施工时间：2010.7—2013.3
摄影师：©Tim Hursley (courtesy of the architect)

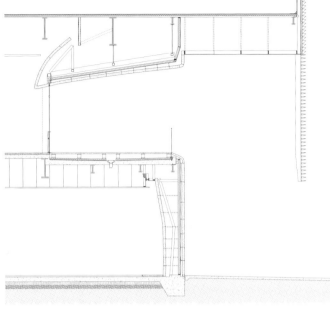

a-a' 剖面图 section a-a'

百叶窗/立面详图选项
louver / facade detailing option

15.2cm百叶窗; 30.5cm间隔
6" louver; 12" spacing

7.6cm百叶窗; 15.2cm间隔
3" louver; 6" spacing

7.6cm百叶窗; 7.6cm间隔
3" louver; 3" spacing

15.2cm百叶窗; 交错间隔
6" louver; staggered spacing

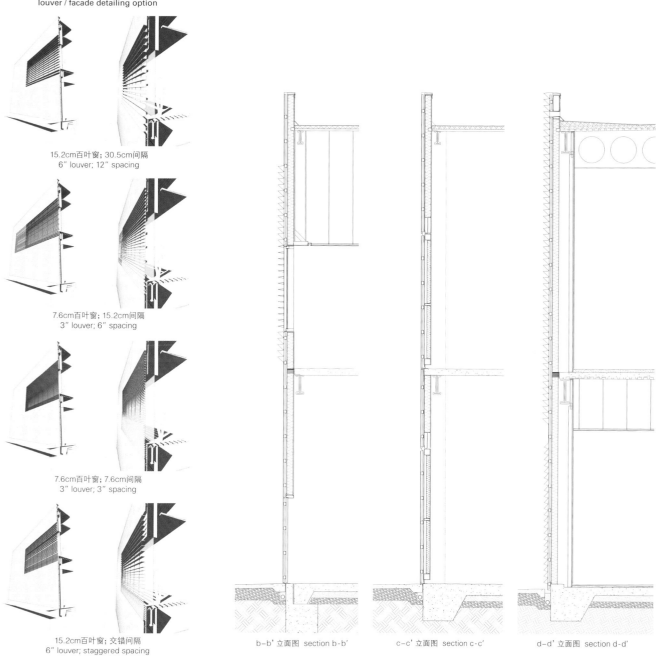

b-b' 立面图 section b-b'   c-c' 立面图 section c-c'   d-d' 立面图 section d-d'

The resolution is, first, to interpret athletics as a component of cultural history rather than as independent themes. While sports and regional history may appeal to different audiences, the exhibits and configuration explore interconnections between the two. The spaces flow visually and physically together, configured to accommodate state-of-the-art exhibits, education and support functions. Visitors however can experience both narratives either separately or simultaneously.

Second, historical pastiche is set aside in favor of a design language in response to the site. The internal organization is an extension of the existing meandering urban circulation, while the design mediates the scale and character of the historic commercial core and adjacent residential neighborhood. The "simple" exterior, clad with pleated copper panels, alluding to the shutters and clapboards of nearby plantations, contrasts with and complements the curvaceous interior within. The louvered skin controls light, views and ventilation, animates the facade, and employs surface articulation previously achieved by architectural ornamentation. The flowing interior emerges at the entrance, enticing visitors to walk into the evocative exhibit spaces within.

Third the design reflects the carving of the ancient river whose fluvial geomorphology inspired the dynamic interior form. The dynamic foyer is sculpted out of 1,100 cast stone panels, seamlessly integrating all systems and washed with natural light from above. The white stone references bousillage, the historic horse hair, earth and Spanish moss utilized by 17th century settlers.The flowing surfaces reach into the galleries, serving as "screens" for film and display. At the climax of the upper level, the path arrives at a veranda overlooking the city square, sheltered by copper louvers, further connecting the interior to the public realm.

Trahan Architects

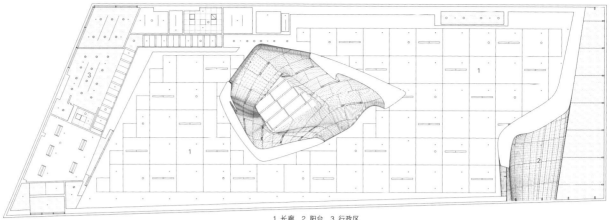

1 长廊 2 阳台 3 行政区
1. gallery 2. veranda 3. administration
二层_反射天花板 second floor _ reflected ceiling

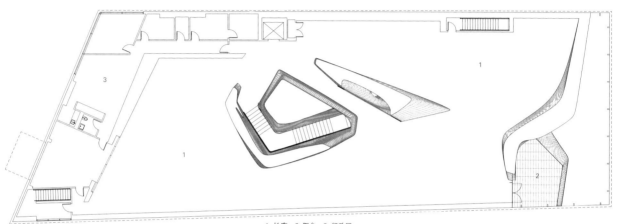

1 长廊 2 阳台 3 行政区
1. gallery 2. veranda 3. administration
二层 second floor

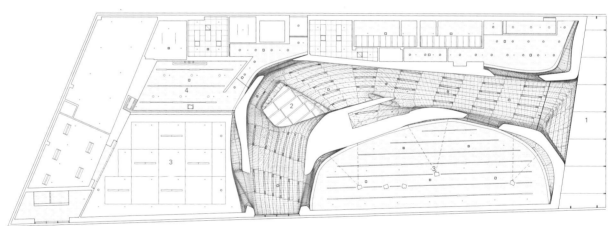

1 门廊 2 门厅 3 长廊 4 教室
1. porch 2. foyer 3. gallery 4. classroom
一层_反射天花板 first floor _ reflected ceiling

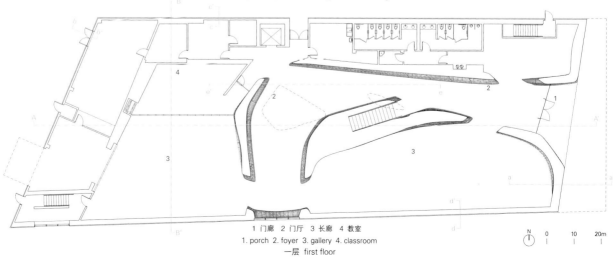

1 门廊 2 门厅 3 长廊 4 教室
1. porch 2. foyer 3. gallery 4. classroom
一层 first floor

球面组装进程
global assembly process

1 原有的接头和砌缝
1. inherited head and bed joints

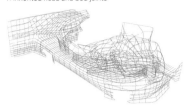

2 通过Catia软件来加厚嵌板
2. thicken panels via Catia

3 成形表面的钢支架
3. generated shaped surface's support steel

4 嵌板连接点
4. generated panel connections

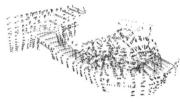

5 符合要求的细部系统
5. complied detailed systems

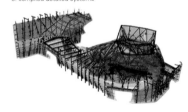

6 完全嵌装在钢基座上的系统
6. fully nested systems in steel base

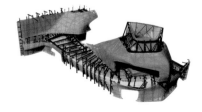

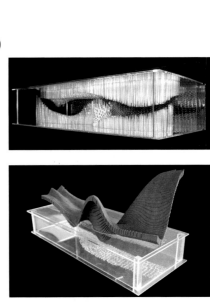

连续的表面
continuous surface

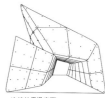

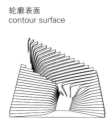
轮廓表面
contour surface

给曲面镶入嵌板
paneling surface curvature

连续的平滑表面
continuous smooth surface

带状平面的均匀分布
even distribution of planar ribbons

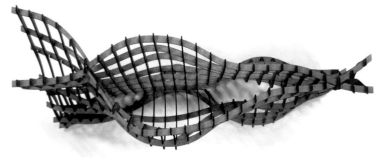

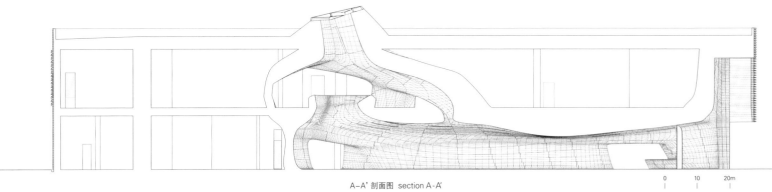

A-A' 剖面图 section A-A'

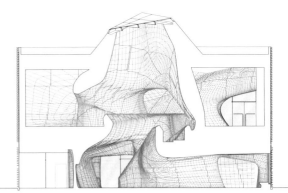

B-B' 剖面图 section B-B'

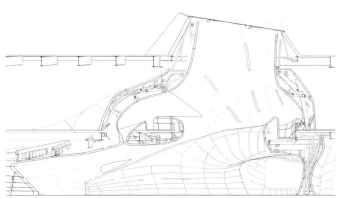

e-e' 剖面图 section e-e'

波兰犹太人历史博物馆
Lahdelma & Mahlamäki Architects

华沙一直是犹太人最重要的城市之一,在第二次世界大战之前,有五十万犹太居民居住在此。

博物馆的场地距老华沙中心一公里。该场址是一个公园,周围环绕着住宅楼宇。今天坐落于此的公园由德意志联邦共和国前总理维利·勃兰特命名,是华沙犹太人区的核心位置。公园里矗立着华沙犹太区起义纪念碑。

博物馆旁边的华沙犹太区起义纪念碑是设计的出发点之一。博物馆和纪念碑前的广场紧密连接,相得益彰,形成一处新的城市空间。主入口位于纪念碑一侧,在此处,利用通向景观区的小桥形成了一系列连续空间。建筑的基本形式简单紧凑,减少了对周围公园区域的影响。竞标评审团对它的评价是"没有不必要的修辞,简洁且优雅"。

竞标提案给它命名为"Yum Suf",象征性地指代主厅的建筑。空间的设计灵感来自《旧约》的传说,同时大厅的形式让人想起普遍的、抽象的自然现象。主厅是建筑结构最重要的因素,纯净无声地把博物馆介绍给游客。

博物馆是一个多功能中心,可用于与犹太传统相关的研究、展览、教育和文化。展览馆的核心区设有一个大厅,类似一处半完工状态的空间,占地5000m²,位于主大厅的下面。该展览呈现从中世纪到今天的犹太文化的不同形态和时期,大屠杀只是展览的一个主题。展览以叙事的方法传递讯息,使用了重建、移动图片和各种人造场景的方法,还有少量的历史物件。核心展厅的平面是竞赛源材料的一部分,同时建筑设计中考虑了功能要求和展览的画廊划分。

博物馆特别关注年轻人,预计每年接待访客50万人。

建筑的框架是现浇混凝土。框架系统的一部分为自由形态的墙壁和与之相连的曲面天花板。钢结构和混凝土喷墙的总厚度约为60cm。作为承重结构的弧形墙的设计特别具有挑战性。据建筑师所知,这是已被实现的最大的、均匀的几何双曲面建筑。设计师专门为这个项目开发

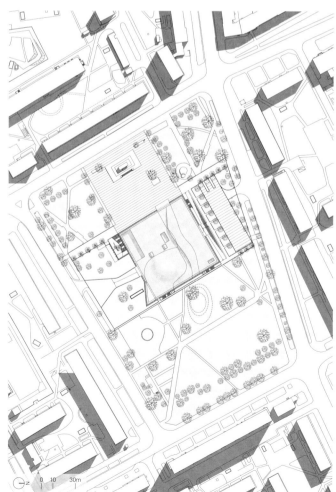

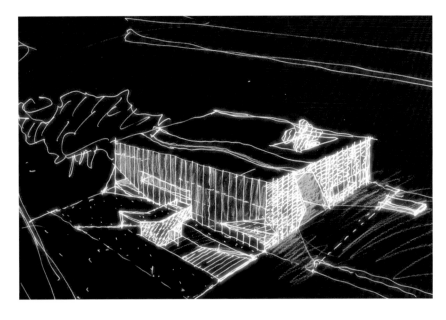
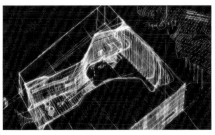

了软件，帮助Markus Wikar和Lahdelma & Mahlamäki的建筑师设计部分结构。

双层立面的外层覆有层压玻璃面板和经过预处理的多孔铜板。

The Museum of the History of Polish Jews

Warsaw has been one of the most important cities for Jews; before the Second World War, there were half a million Jewish inhabitants in the city.

The museum's plot is about a kilometer away from the old center of Warsaw. The site is a park surrounded by residential buildings. The park located there today, named after the former Chancellor of the Federal Republic of Germany, Willy Brandt, formed the core of the Warsaw ghetto. The Memorial to the Warsaw Ghetto Uprising is erected in the park.

The Memorial to the Warsaw Ghetto Uprising located next to the museum is one of the key points of departure for the design. The square in front of the memorial and the museum are sufficiently close together and their dimensions are mutually compatible: the square and the museum building will form a new urban space. The main entrance is placed on the memorial side of the building, from which a series of spaces continues via a bridge towards the landscape. The basic form of the building is compact and simple, reducing its footprint in the surrounding park. The competition jury stated in its evaluation, that the concept had been realized "without unnecessary rhetoric, with simplicity and elegance".

The name of the competition proposal, "Yum Suf", symbolically refers to the architecture of the main hall. The inspiration for the space has been the legends of the *Old Testament*, although at the same time forms of the hall refer to the universal and abstract phenomena of nature. The main hall is the most important element in the architecture of the building: a pure and silent space introducing the museum to the visitors.

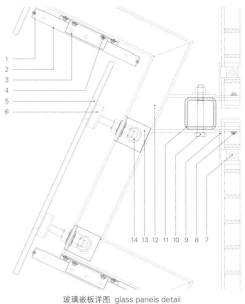

玻璃嵌板详图 glass panels detail

1. copper rivet ø3.2x8
2. perforated copper sheet 1mm predia 406-000
3. L25x25x2 AL
4. drilling with sealing screwed HIL TI 51Z S-MD
5. Gr. glasses; by static requirements; min 10m
6. glass fitted
7. cylinder screw M8x70-A2 + washer + nut
8. self-drilling screw with gasket; HIL TI 51Z S-MD
9. L60x30x5 steel
10. screw M12x90-A4 + washer + nut
11. RK60x60x4 steel
12. RP100x60x4 steel
13. L75x55x6 steel L = 50
14. screw M14x25-A4

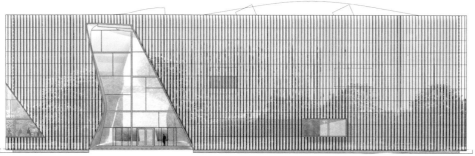

东北立面 north-east elevation

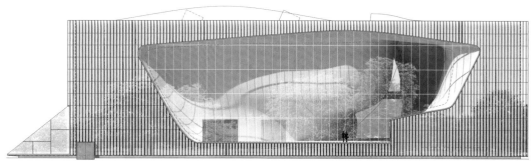

西南立面 south-west elevation

项目名称：The Museum of the History of Polish Jews
地点：6 Anielewicza St., 00-157 Warsaw, Poland
建筑师：Architects Lahdelma & Mahlamäki Ltd., Kurylowicz & Associates in Warsaw
作者：Rainer Mahlamäki
总负责人：Riitta Id(design phase), Maritta Kukkonen
合作建筑师：Jukka Savolainen, Mirja Sillanpää(built-in furniture), Miguel Freitas Silva, Markus Wikar(master model, geometry and scripting)
波兰合作建筑师：Kurylowicz & Associates; Stefan Kurylowicz(until 2011), Ewa Kurylowicz(2011~2013), Pawel Grodzicki(design phase), Marcin Ferenc, Tomasz Kopeć, Michal Gratkowski
办公室的家具设计师：Grupa Plus Architecture Studio
结构工程师：Pol-Con Consulting／电气工程师：Elektroprojekt SA
甲方：City of Warsaw and Ministry of Culture／承包商：Polimex-Mostostal SA
功能：museum consisting of exhibition spaces, auditorium, offices
用地面积：12,442m²／总建筑面积：4,400m²／有效楼层面积：18,300m²
材料：principal materials_silk printed glass, copper, concrete／principal structure_concrete, steel
造价：USD 48 million
施工时间：2009.7—2013.5／竣工时间：2013
摄影师：
©Juho Haiminen (courtesy of the architect) - p.115, p.118, p.123
©photoroom.pl (courtesy of the architect) - p.117 bottom-right, p.121 top-left, bottom
©Wojciech Krynski (courtesy of the architect) - p.114 bottom, p.121 top-right
©Pawel Paniczko - p.112~113, p.116~117, p.122

The museum building is a multi-functional center for research, exhibition, education and culture relating to the Jewish heritage. The core exhibition area comprises a hall that resembles an only half-finished space, almost five thousand square meters in area, beneath the main lobby. The exhibition will present the different forms and periods of Jewish culture from the Middle Ages on until today – the holocaust is only one of the themes of the exhibition. The message will be conveyed in the form of a narrative exhibition, making use of reconstruction, moving pictures, and various constructed milieus and also minor extent historical objects. The plan for the core exhibition was part of the competition source material, and the functional requirements and gallery division for the exhibition were to be considered in designing the building. There is to be a special focus on young people. The museum expects half a million visitors annually.

The frame of the building is cast-in-situ concrete. The free-form walls and the curving shapes of the roof connecting to them form part of the frame system. The total thickness of the steel structured and sprayed concreted wall is about 60cm. The design of the curved walls, which are bearing structures, was particularly challenging. To the best of the architects' knowledge it is the biggest uniform, geometrically double-curving surface that has ever been realized. The design was partly implemented with assistance of software developed by the designer specifically for this project by Markus Wikar and Architects Lahdelma & Mahlamäki.

The outer layer of the double facade is to be clad with laminated glass panels and pre-treated, perforated copper panels.

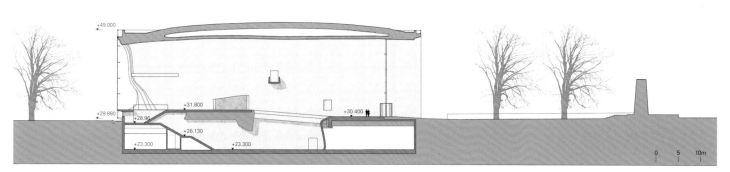

A-A' 剖面图 section A-A'

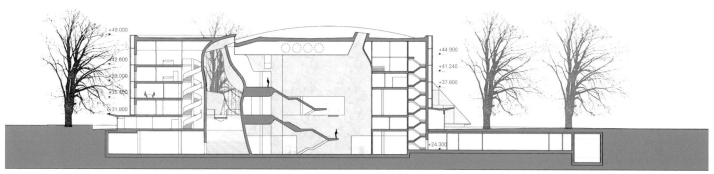

B-B' 剖面图 section B-B'

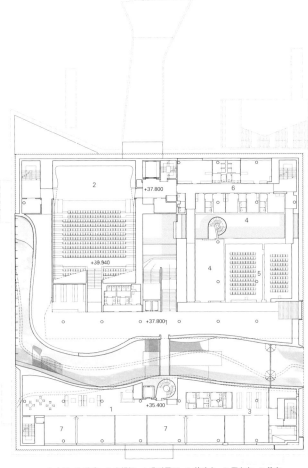

1 大厅 2 礼堂 3 衣帽间 4 临时展厅 5 放映室 6 更衣室 7 教室
1. lobby 2. auditorium 3. cloakroom 4. temporary exhibition
5. projection room 6. dressing room 7. classroom
二层 second floor

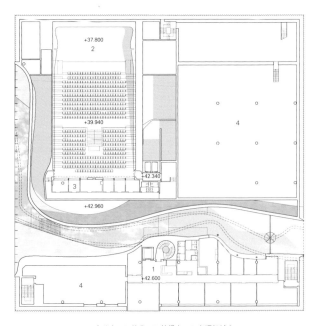

1 办公室 2 礼堂 3 转播室 4 空调机械室
1. office 2. auditorium 3. translator room 4. air conditioning machine room
四层 fourth floor

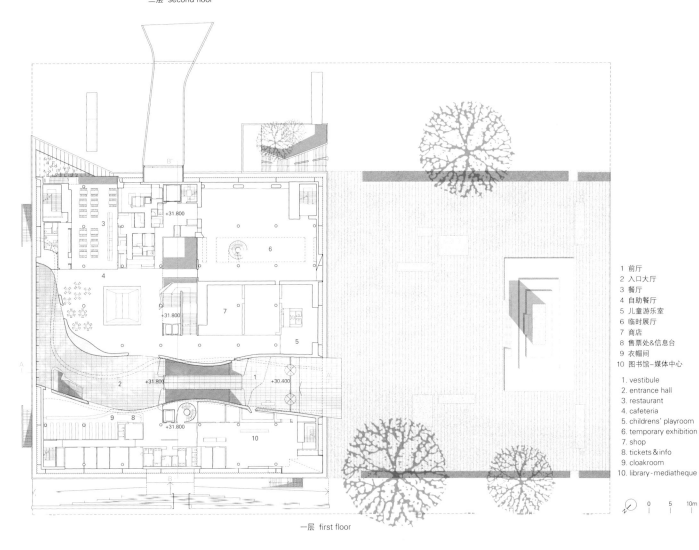

1 前厅
2 入口大厅
3 餐厅
4 自助餐厅
5 儿童游乐室
6 临时展厅
7 商店
8 售票处&信息台
9 衣帽间
10 图书馆-媒体中心

1. vestibule
2. entrance hall
3. restaurant
4. cafeteria
5. childrens' playroom
6. temporary exhibition
7. shop
8. tickets & info
9. cloakroom
10. library - mediatheque

一层 first floor

拉梅骨多功能亭台

Barbosa & Guimarães Arquitectos

绿树葱葱的圣埃斯特旺山公园俯瞰整座城市，通过Our lady of Remedies教堂庄严的楼梯（沿着菩提树大街延伸），一路连到拉梅骨市的中心。

与18世纪雄伟的综合性建筑一起位于公园脚下，多用途的亭台使圣埃斯特旺的山坡变得柔和，通过一个广场和设在屋顶的圆形剧场，并且利用自然的梯度来弱化其庞大的体积。

当前的集市空间成为新亭的前厅，十分不协调。因此其范围被重新设计，使其变成更能控制的空间，与周边街道建立了一种新的关系。

主要入口都位于南立面，那里的大剧场连接着一高（地上较高的位置）和一低（较低的位置）两个广场。

新广场与下面的亭台在北端融合，令城市公园获得了一个新的方位。

三处新的空间，公园、集市和广场，径直连接到Our lady of Remedies教堂的林荫路和楼梯共同构成的城市轴线，显著地增强和提升了拉梅骨市的公共空间。

作为嵌入的、锚固的凉亭设施，具有多种用途，适合用作多用途竞场场和门厅，其跨度为50m，篷高为10m。该馆还提供淋浴设施、更衣室、多功能房间和可容纳120人的礼堂，更加凸显了它的通用性。亭台有一个地下四层的停车场，连接了高高低低的街道。

拉梅骨市的建筑和地基都使用了花岗岩，为新公共空间披上了一件新的外衣，强化了项目所追求的连续性和融合性的特点。

Lamego Multipurpose Pavilion

Overlooking the entire city, the leafy park of Mount Santo Estêvão is directly related to the center of the city of Lamego through the imposing Stairway of the Sanctuary of Our Lady of Remedies which threads its way along the Lime Tree Avenue.

Located at the foot of the park, together with the monumental 18th century complex, the multi-purpose pavilion mellows the hillside of Mount Santo Estêvão, taking advantage of the natural gradient to nullify its volume, through a plaza and an amphitheater installed on its roof.

The current fair space, which now acts as an anteroom for the new pavilion, was out of character. Its limits were redesigned, transforming itself into a more controlled space, establishing a new relationship with the surrounding streets.

On the south elevation, where the main accesses to the building are located, the great amphitheater allows a connection to be made between the two plazas, the new one at a high level above

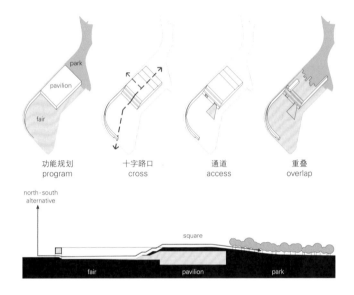

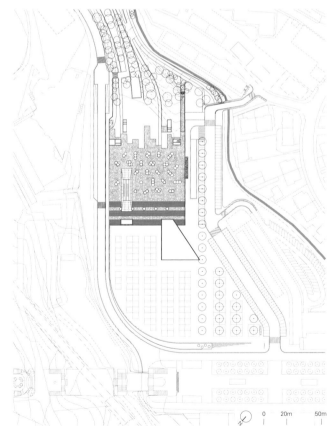

ground and the other at a low level.

The urban park gains a new orientation, the new plaza over the pavilion merging at its northern end.

These three new spaces, park, fair and plaza, in direct conjunction with the urban axis defined by the Alameda and Stairway of the Sanctuary of Our Lady of Remedies, significantly reinforce and upgrade the public space of the city of Lamego.

The pavilion, the anchoring facility in the intervention, allows various uses, the fruit of the multipurpose nature of the arena and of the foyer, which their 50-meter span and ceiling height of 10-meter make possible. The pavilion also offers shower facilities and changing rooms, a multipurpose room and auditorium for 120 people which complement its versatility. With the pavilion there will be a car park, with four underground floors, allowing streets with high and low levels to be connected.

Granite, present in the subsoil and in the architecture of Lamego, coats the new public spaces, strengthening the character of continuity and integration that the project seeks.

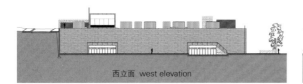
西立面 west elevation

南立面 south elevation

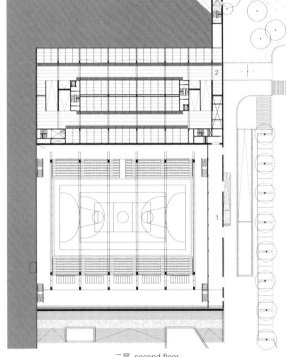

1 技术区域
2 停车场
1. technical area
2. parking

二层 second floor

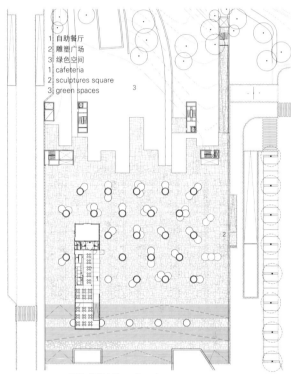

1 自助餐厅
2 雕塑广场
3 绿色空间
1. cafeteria
2. sculptures square
3. green spaces

屋顶_雕塑广场 roof_sculptures square

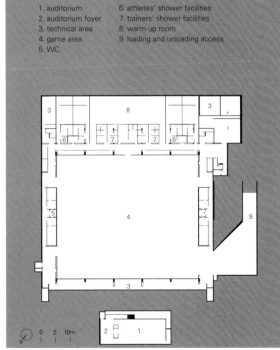

1 礼堂
2 礼堂门厅
3 技术区域
4 游戏区
5 卫生间
6 运动员淋浴区
7 教练员淋浴区
8 热身室
9 装卸通道

1. auditorium
2. auditorium foyer
3. technical area
4. game area
5. WC
6. athletes' shower facilities
7. trainers' shower facilities
8. warm-up room
9. loading and unloading access

地下一层 first floor below ground

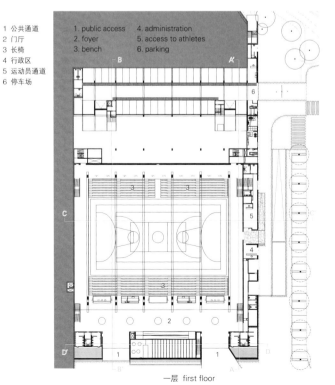

1 公共通道
2 门厅
3 长椅
4 行政区
5 运动员通道
6 停车场

1. public access
2. foyer
3. bench
4. administration
5. access to athletes
6. parking

一层 first floor

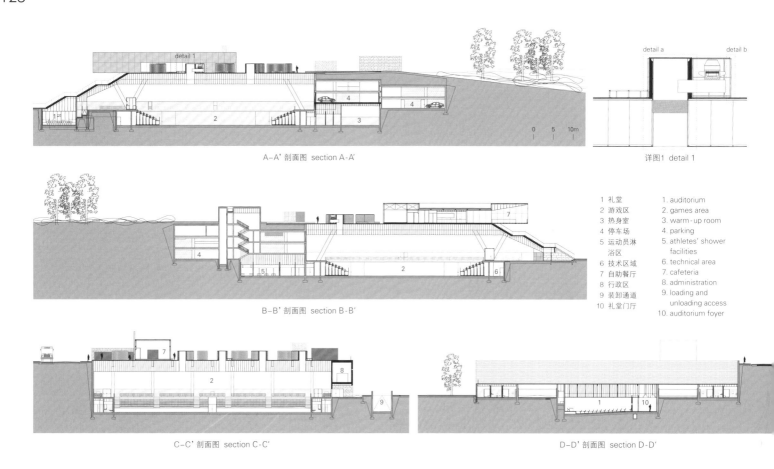

A-A' 剖面图 section A-A'   详图1 detail 1

B-B' 剖面图 section B-B'

C-C' 剖面图 section C-C'   D-D' 剖面图 section D-D'

1 礼堂	1. auditorium
2 游戏区	2. games area
3 热身室	3. warm-up room
4 停车场	4. parking
5 运动员淋浴区	5. athletes' shower facilities
6 技术区域	6. technical area
7 自助餐厅	7. cafeteria
8 行政区	8. administration
9 装卸通道	9. loading and unloading access
10 礼堂门厅	10. auditorium foyer

1. reinforced concrete tube 2,000x2,000x192mm
2. flint coat painting
3. corten steel sheet 1.5mm
4. iron tubing 30x30x3mm
5. corten steel sheet 5mm
6. iron tubing 50x50x3mm
7. led for white light illumination
8. bituminous screens
9. granite 30mm
10. mortar fixative 5mm
11. wood 10mm

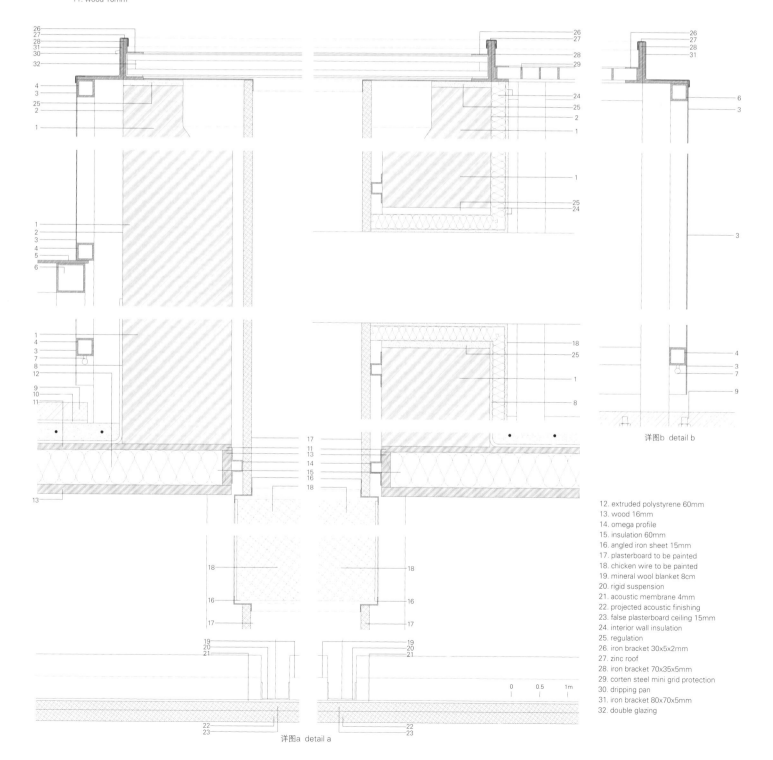

12. extruded polystyrene 60mm
13. wood 16mm
14. omega profile
15. insulation 60mm
16. angled iron sheet 15mm
17. plasterboard to be painted
18. chicken wire to be painted
19. mineral wool blanket 8cm
20. rigid suspension
21. acoustic membrane 4mm
22. projected acoustic finishing
23. false plasterboard ceiling 15mm
24. interior wall insulation
25. regulation
26. iron bracket 30x5x2mm
27. zinc roof
28. iron bracket 70x35x5mm
29. corten steel mini grid protection
30. dripping pan
31. iron bracket 80x70x5mm
32. double glazing

详图a detail a

详图b detail b

项目名称：Lamego Multipurpose Pavilion
地点：Lamego, Portugal
建筑师：José António Barbosa, Pedro Lopes Guimarães
合作建筑师：Miguel Pimenta, Paula Fonseca, José Marques, Henrique Dias, Filipe Secca, Mafalda Santiago, Raul Andrade, Pablo Rebelo, Paulo Lima, Ana Campante, Ana Carvalho, Ana Mota, Daniela Teixeira, Eunice Lopes, Nuno Felgar, Rui Grangeio
项目主管：José Cruz / 结构工程师：José Ferraz / 水利工程师：Pedro
电气工程师：Joaquim Silva / 水暖空调和煤气工程师：Pedro Albuquerque
施工单位：Irmãos Cavaco, Gabriel a.s. couto, Francisco Pereira Marinho & Irmãos s.a.
甲方：Lamego Renova
用地面积：25,654m² / 总建筑面积：5,200m² / 有效楼层面积：10,454m²
竞赛时间：2006 / 项目设计时间：2006—2008 / 施工时间：2009—2011
摄影师：©José Campos (courtesy of the architect)

曾经的工业建筑向文化建筑的转变
Cultural Shift from

我们见证了遗留工业建筑转变成文化机构和旅游胜地后所获取的日益增加的巨大效益。特别是近十年来，过去用于生产活动的建筑通过注入文化元素找到了能够使它们复兴的方法。但遗留的工业基础设施为何与"文化"结合得如此完美呢？一栋建筑在由工业生产转化至文化产出时涉及到哪些因素呢？既然涉及到要保护的形式，那么具体有什么需要被保护呢？是建筑结构、地理位置、该地的文化意蕴还是现在与过去之间的碰撞？

本章展示的项目充分说明了建筑师对后工业景观持有的多种态度，但同时也强调在由工业生产到文化产出转变过程中涉及的诸多冲突。文章通过一系列话题和相互交织的叙述审视了这一现象。同时，我们需要以一种特定的心境来倡导保护欲，它使我们思考时间与地点的关系问题，并最终形成一处不仅仅歌颂过去的环境。重建的工业场地真正地为建筑的高级化进程做出了贡献，但也被用作为投机开发的营销工具。虽然如此，在应用到文化空间的过程中，这些工业空间所显现出的优异质量也留下了深深的印记。对这些建筑的再利用开创了一个先例并且生成如今被转变成新建文化机构或扩建的建筑模型，即应用建筑语言且追忆工业遗产的模型。

We have witnessed a great wealth of cultural institutions and destinations rising out of leftover industrial buildings. Especially over the last decade, architecture from past manufacturing activities found their recovery through injections of culture. Why do leftover industrial infrastructures and "culture" marry so well? What is involved when a building shifts from industrial production to cultural production? And since it concerns a form of preservation, what is it that is being preserved? The building fabric, the location, the cultural significance of the place or the encounter between present and past?

The projects presented within this chapter provide a considerable illustration of the manifold attitudes taken towards post-industrial landscapes, but also highlight a number of clashes involving their shift from industrial to cultural production. The text examines this phenomenon through a series of topics and intertwined stories. Whilst it needed a certain frame of mind to initiate a thirst to preserve, it made us question time in relation to place, and eventually create environments that surpass a mere celebration of the past. Recovered industrial sites have been genuine contributors to processes of gentrification, but have also been "used" as marketing tools for speculative developments. Nevertheless, the qualities these industrial spaces presented in the process of being adopted to cultural spaces definitely left a profound mark. Their re-use established a precedent and generated architectural models that are now been transferred to newly built cultural institutions or extension, models that are adopting an architectural language reminiscent of this very industrial heritage.

建在旧储气罐内的多功能大厅_Multipurpose Hall Fitted in Former Gasholder/AP Atelier
C-Mine建筑_C-Mine/51N4E
胶印厂的改造_Rebirth of the Offset Printing Factory/Origin Architect
人人剧院_The Everyman Theatre/Haworth Tompkins Architects

从工业到文化，人员流动取代商品流通_From Industrial to Cultural,
Substituting a Flow of Goods with a Flow of People/Tom Van Malderen

bygone Industry

保护欲

20世纪中叶,许多重要的工业遗留建筑被拆除。50年代和60年代泛滥的无良城市更新政策毁坏了工业城市的很多部分。尽管怀着最大的善意和留下的些许希望,但是这种摧毁行动还是没有达到当初预想的社会和经济目标。

随之而来的醒悟和沮丧形成了塑造当今工业遗产观念的不切实际的势头,即改造理论的发展以及对这些工业遗产的保护欲。虽然过程比较缓慢,但城市的确作为一个"整体"被人们认为是我们共同的文化遗产,也包括过去几百年,甚至几十年里增建的建筑物。如今,我们对城市的看法有所改变,我们将它们看作是表现过去、现在和将来的文化实体。

最近一段时间,我们见证了越来越多的工业遗产正在申请加入联合国教科文组织官方世界遗产名录,同时更多后工业场地正排队等待认证。其实,这种不断升级的关注直接指向某些建筑师和规划师所提到的过度保护的状态。然而发生了逆转,过度保护可能会给我们文化遗产带来一些问题,它会使这座城市停滞在"过去",从而失去"未来"。

从边缘到中心

工业场地魅力的一部分存在于它们坐落的位置上。在规划和建造时,大多数场地被安排在城郊地区,远离生活区。随着多年的城市扩张和蔓延,如今这些场地占据着许多当代城市的核心位置,使它们转变成充满机会和具备潜在价值的地方:作为临近人口稠密区域的地方,这些地点通常交通便利并且可以随时迎接大量的参观者。

由原地建筑设计有限公司开展的胶印厂的改造项目很好地阐述了这一点。这座曾经繁荣兴盛的北京胶印厂分别建造于20世纪60年代、70年代和90年代,已经由一个工业活动中心转变成一个废弃仓库的集散地,位于民宅和城市有名的胡同和小巷的边缘。建筑师以人员流动替代了该地方最初规划形成的物流。临近的街道穿过原始的建筑物,盘旋的楼梯和浮桥结合起来,连接现有的后院和翻新建筑(经翻新带有新建庭院、空中花园和延伸整个建筑屋顶的露台)。

A Thirst to Preserve

During the middle of the twentieth century a lot of significant industrial leftovers got wiped out. The infamous urban renewal policies dominated the 1950's and 1960's destroyed numerous sections of industrial cities. Despite the best intentions and a few exceptions left aside, this wipe-out didn't achieve the social and economic goals it had set out for.

The disillusion and disappointment that followed created the ideal momentum to shape today's industrial heritage concept, the development of reclamation theories and a thirst to preserve. Slowly but certainly, the city as a "whole" got acknowledged as our collective heritage, including structures that were added only over the past hundred years, or even decades. Our perception of the city altered, and today, cities are looked upon as cultural entities that contain representations from the past, via the present, to the future. In recent time, we have witnessed an increasing volume of our industrial heritage making it on to the official UNESCO World Heritage List, whilst many more post-industrial sites are lining up for recognition. In fact, such an escalating attention is directed towards preservation that certain architects and planners start referring to as a situation of over-preservation. As a reversal of fortune, over-preservation might turn into a problem for our collective heritage, it might freeze "the past" and take "the future" out of the city.

From the Periphery to the Centre

Part of the allure of industrial sites is to be found in their location. At the time they were planned and erected, the majority of them were considered to be suburban sites, often disconnected from living areas. Following years of urban expansion and sprawl these sites now occupy central spots in the middle of many contemporary cities, turning them into places of opportunity and latent value: places in close proximity of densely populated areas, generally well connected and ready to reach large crowds of visitors.

The Refurbishment of the Offset Printing Factory by Origin Architect illustrates this very well. Built incrementally through the 1960s, 1970s and 1990s, this once-thriving Beijing Offset Printing Factory had changed from being a centre of industrial activity to a collection of desolate warehouses, hemmed in amongst the homes and city's famous hutongs or alleyways. The architects literally exchanged the flow of goods, for which this place had been devised, with a flow of people. The neighbourhood's streets are

模糊地段和我们对遗迹的怀旧渴望

先前废弃不用的生产活动场地具有空置性、潜在性的和开发性的特征——被Ignasi de Solà-Morales设计为模糊地段。建筑未被使用的状态使人产生一种自由的感觉,同时体现出一处充满希望甚至是期待的场地。通过项目本身和投资,当今的建筑和城市设计采取的最为普遍的方法是将这些地方和建筑重新融入城市空间的生产网,因此使它们成为高效的、繁忙的且实用的城市的一部分。De Solà-Morales提醒我们在循环再利用这些模糊地段时,应该尝试去保留存在于空置和未被使用场地中的价值。这些模糊地段是具有特殊地位的场地,是现在和过去产生碰撞的地方。同时,他还指出19世纪城市文化中以同样方式开发的城市公园是一种对新型工业城市的回应和对策。我们的后工业文化需要的是自由的、未界定的和非生产性的空间。这一次,它与大自然的神秘概念无关,而与回忆的经历有关。活跃于当代文化中的另一组情感是对真实性的向往,以及对现代遗迹的怀旧渴望的诠释。引发怀旧情绪是因为"遗迹"似乎带有一种已然从我们自己年代消失的希望,它使我们向往先前几十年假定的理想和明确性。Andreas Huyssen将这种保留废弃工业场所的渴望描述为"记忆文化"和当前对西方历史痴迷的一部分,它是对加速现代化的一种反应;是对打破日常空闲空间漩涡的一种尝试并且主张时间和回忆观念。但是我们该如何留住周围所有老旧的事物呢?建筑方面的循环再利用政策能否会有效地唤醒回忆,还是会像De Solà-Morales的改造一样,实际上具有消除这些记忆的风险?

调整历史文物

本章中显示的一些建筑只是侥幸逃过被拆除的命运。它们幸免于难,然而经济事实在改变,技术进步了,并且持续的城市更新计划也在发起。尽管有模糊地带更加浪漫化的概念,但是当这些生产使用的建筑失去它们的使用价值时,疏于照管的状况同样使人不安,因为它们的自然破败不断地提醒着我们后工业化存在的问题。在很多情况下,在这些建筑最初投入运行时,城市居民并没能完全调整自身以适应失业和相关生活方式的改变,而大型工业设施的关闭遗留下的是经济的弊端和整个工作环境的消失。而这正是AP工作室设计的储气罐多功能大厅和51N4E设计的C-Mine建筑的场地形成的状况。它们都是宽阔的工业园区的一部

filtering through the original structures and a combination of hovering staircases and floating bridges connects existing backyards and refurbished buildings with newly created yards, hanging gardens and terraces stretched all over the buildings' rooftops.

Terrain Vague and our Lust for Ruins

Disused sites of previous production possess the allure of vacancy, potential and openness – coined by Ignasi de Solà-Morales as terrain vague. The absence of use evokes a sense of freedom, and at the same time embodies a space of promise and even expectation. Nowadays' approach most commonly adopted by architecture and urban design, by means of projects and investment, is to reintegrate these spaces and buildings into the productive mesh of urban spaces, thus making them part of the efficient, busy and effective city. De Solà-Morales reminds us that when recycling these terrains vagues, we should try to retain the values residing in their vacancy and absence. These terrains vague are privileged sites of identity, of encounters between present and past. He also points out that in the same way the invention of the city park was developed within the nineteenth-century urban culture as a response and antidote to the new industrial city, and our post-industrial culture calls for spaces of freedom, that are undefined and unproductive. This time round, is not associated now with the mythical notion of nature, but with the experience of memory. Another set of sentiments that is active within contemporary culture is the longing for authenticity, and its translation into a nostalgic lust for the ruins of modernity. Nostalgia is being triggered because "the ruin" seems to hold a promise that has vanished from our own age, and makes us yearn for the assumed ideals and clarity of previous decades. Andreas Huyssen describes this desire to keep abandoned industrial sites as part of "memorial culture" and the current western fascination with history, as a reaction to the accelerated speed of modernization; as an attempt to break out of the swirling of empty space of the everyday present and to claim a sense of time and memory. But how do we go about holding on to all the old and ageing stuff surrounding us? Will recycling strategies for architecture effectively evoke memories, or actually risk erasing them, as De Solà-Morales alerted?

Reconciling Reminders of the Past

Some of the buildings shown in this chapter only narrowly escaped their demolition. They survived, whilst economic realities were changing, technological advances were made, and ongoing urban renewal schemes instigated. Despite the more romanticizing notion of the terrain vague, when these manufacturing buildings lost their use, the condition of neglect is equally unsettling because their physical decay represents a constant reminder of the problems of post-industrialization. On many occasions, city

分，而它们的实际存在是对曾经支配整个地区活动和文化的一种外在的提醒。该城镇中曾经在此工作的居民如今是文化活动的参观者。人们仍然会聚集于此，从这一层面上来说，该场地实现了循环再利用的目的。然而，更重要的一点是，它们的循环再利用需求不仅仅体现对过去的缅怀，还能在重建未来方面发挥积极作用。对这一共同遗产的翻新势必会形成一种共同的地区认同感和开发新型多功能景观的机遇，以确保未来持续的复原和改造。C-Mine建筑将一个先前的矿业综合设施的发电所改造成为一个地区文化中心。通过大量重塑该建筑的后工业环境，51N4E的建筑师没有使"传统工业"和"新文化"仅仅是相互毗邻，而是在两者之间建立了一种更加紧凑并且空间上能够积极交流的环境。如此一来，现在、过去和将来的界限变得模糊起来。

清洁时代的遗留建筑

仔细观察大多数工业建筑的物理表现，我们可以看出这些大型开放空间的优势在于允许充足的自然阳光的进入，在适用新建筑规划方面也具有极大的灵活性。多年来，旧厂房和仓库已然成为艺术家们钟爱的工作/生活空间。因此，循环再造的工业空间开始与艺术创作和创造性产生关联。许多新博物馆遵循这一概念并且越来越喜欢在与创作艺术品空间相似的地方展出作品。虽然这些充满灰尘、污渍和剩余材料（经常被浪漫化）的废弃空间非常适合艺术家的形象；而在另一方面，文化机构却需要遵从当代的我们对清洁度的要求。我们不希望在博物馆——一处小心保存物品并且使用条件苛刻的地方看到灰尘。我们可以将这种情况说成是原有环境和新环境之间的冲突。那么"清扫"这些空间应该采取什么策略呢？我们如何保证既能保护空间又能实现清洁目的呢？例如，当表面擦洗干净，或保留层层的污垢和饱经风霜的材料意味着什么呢？我们可以在本章展示的项目作品中看出建筑师对先前工业作业的物质实体所持有的不同的潜在态度。尤其是在储气罐多功能大厅和C-Mine项目中，我们可以看到新创建的空间与先前的砂砾表面和机器文物共存，形成材料、时期和功能方面更加紧密的并置。在北京胶印厂项目中，我们看到清扫过去和披露过去两种元素的并存。一些较新的扩建物被移除以形成户外空间，同时将文化大革命时期的砖墙裸露出来，并分别结合不同年代的嵌入结构。

residents have not fully reconciled themselves with the loss of employment and the associated way of life they were running when these buildings were originally operating. In the case of sizeable industrial facilities, the closing leaves alongside an economic drawback, the loss of an entire working culture. This was definitely the case for the sites where the Gasholder Multipurpose Hall by AP Atelier and C-Mine by 51N4E were created. They formed both part of extensive industrial complexes, and their physical presences are strange reminders of the activity and culture that once dominated the entire region. The town's population that once used to work there, are today's visitors to the cultural events. The sites still bring people together, and in that sense, have been reclaimed. However, and more importantly, their reclamation needs to represent more than the celebration of the past, and become an active part in reconstructing the future. It is imperative that the rehabilitation of this shared patrimony contributes to the creation of a shared local identity and the opportunity to develop new multi-functional landscapes that ensure an ongoing recovery and transformation for the future. The C-Mine project transformed the powerhouse of a former mining complex into a regional cultural centre. Through an intense reimagining of its post-industrial context, the architects 51N4E avoided the mere adjacency of "heritage industrial" and "new culture", but established a more compressed and spatially active exchange between the two. The past and the future get blurred in the now.

Leftovers in the Age of Cleanliness

Taking a closer look at the physical manifestation of most industrial buildings, we can see the advantage of large open spaces with plenty of natural light that can cater for great flexibility in adjusting to a new building program. For many years now, old factories and warehouses have become a favourite work/living place for artists. As a consequence, recycled industrial spaces started to be associated with the image of artistic production and creativity. Many new museums followed this "image" and grew a liking to display art in a space similar to the one it was produced within. Whilst these abandoned spaces come with an accumulation of dust, stains and leftover materials that suit the – often romanticised – picture of the artist perfectly; cultural institutions, on the other hand, need to abide to our contemporary expectations of cleanliness. We do not expect to find dirt in an art museum, a place where material is carefully conserved and environmental conditions are demanding. We could speak of a clash between the original and the new conditions. What are the strategies taken towards "cleaning out" these spaces? How do we conserve and clean at the same time? What does it mean, for example, when surfaces are scrubbed cleanly, or when layers of grime and weathered materials are retained? We can notice different underlying at-

控制时代的建筑升级

许多转变成文化机构的原有的工业场地为特定历史的发展做出了贡献,它们被称为是对"建筑的升级"。其文化活动支持了当地经济,推动了城市复苏,并且对城市的文化和社会形象做出了贡献。这一点在那些已失去大部分传统工业用途的城市当中体现得尤为明显。像伦敦的泰特现代美术馆这样成功的案例得到了广泛的宣传,并且引起了开始将"自然"的升级进程作为投机和利润丰厚的房地产开发项目工具的开发商的关注。

后工业开发项目的一个难题存在于建筑升级当中,尤其是其中文化发挥的作用,很容易成为一种营销策略而非一项历史性的开发。这些项目很容易成为将"可持续性"和"遗迹保护"术语作为营销标签的、彻头彻尾的投机项目。从这种意义上来说,一连串的问题将接踵而至。当把文化注入作为一种经济策略时,我们可以轻易地看出为何有太多的艺术设施出现,但最终却制约它们走向成功。值得注意的是在许多城市中,文化中心的增长速度超过健康和教育类建筑的增长速度,或与之齐平。将建筑升级作为完成社会工程的明确方案的工具,成为一个更加巨大的风险。这是一种移除或避开通常存在于公共空间中的矛盾或者冲突的手段,以支持一种成功的消费者模式。这种暗淡的场景使我们想到 Theodor Adorno 做出的有关"文化工业"的评论。在每个行业中,其目的在本质上都与经济有关;并且所有的努力都变得注重经济效益。而我们是否应该提醒自己"文化工业"的最大危险在于人们卷入麻木的重复当中,忘记它的存在和影响?建筑升级进程也成就了理查德·佛罗里达的书籍(如《创意阶层的崛起》)的成功和流行。他创作的书籍影响了许多政策制定者,使他们盲目地相信"如果我们建造文化设施,我们将会有更多更加具有创造性的人民"。这种策略引发了各种各样的问号,因为它可以使某些涉及建筑升级的自发和有机过程陷入混乱当中。

工业标准

如果我们以一种不那么悲观的观点来探讨这种由工业向文化的转变,我们可以看到设计工作中一种有趣的演变:工业建筑由遗留物变成具有创造性的循环再利用场地,成为文化机构的普遍环境,并且最终见证其空间和无形的特征被输出至新建的文化场地中。如果我们转向文化

titudes towards the material reality of the previous industrial operations at work in the projects illustrated in this chapter. Especially in the Gasholder Multipurpose Hall project and C-Mine we can see the newly created spaces coexisting with the original gritty surfaces and machine relics, creating a more intense juxtaposition of materials, time periods and functions. In the Offset Printing Factory project in Beijing we see both the element of clearing and exposing of the past. A number of more recent accretions were removed to create outdoor spaces whilst at the same time brick walls from the Cultural Revolution period are now exposed and a series of interventions of different eras are distinctively combined.

Gentrification in the Age of Control

Many former industrial sites that were turned into cultural institutions contributed to a specific historical development referred to as "gentrification". Their cultural activities helped support the local economy, activate urban revitalization, and contribute to the cultural and social image of their city. This was especially the case in cities that have lost much of their traditional industrial uses. Success stories like the Tate Modern in London got widely published and came to the attention of developers who started to employ the "natural" gentrification processes as tools for speculative and lucrative property development.

One of the complications for post-industrial development projects lies within the fact that gentrification, and particularly the role of culture within it, risks becoming a marketing strategy instead of a historical development. Projects risk becoming exclusively speculative, using the terms "sustainability" and "heritage protection" as marketing labels. In this sense, a cluster of hitches can come to the foreground. When the injection of culture is being used as an economic strategy, we can easily see a reason why too many new arts facilities are appearing and limiting their success in the long run. It is noteworthy that in many cities cultural centres grew faster than or on par with buildings in the health and education sectors. An even greater risk is gentrification is being used as a tool to execute an explicit agenda of social engineering. It's a means to remove or bypass the contradictions or conflicts that used to exist in public space, in favour of a successful consumer model. This kind of doom and gloom scenario is a close reminder of Theodor Adorno's critique about the "culture industry" where the aims are – as in every industry - economic in nature; and all endeavours become focused on economic success. And should we remind ourselves that the greatest danger of the "culture industry" is that people, caught up in mind-numbing repetition, forget its presence and its influence? The process of gentrification also got caught up in the success and popularity of Richard Florida's books, such as *The Rise of the Creative Class*. His books influenced many policy makers to a blindfolded belief that "if we have cultur-

产业，工业时代的建筑势必会变成如今的工业标准。法国19世纪经济学家Alfred de Foville阐述到："人类建造房屋，在建造过程中，他们势必会加入一些自己的想法。然而，随着时间的流逝，房屋同样也会使人产生一些想法。"虽说他的阐述适用于他那个年代的住房状况，但是这种对环境和居民相互影响的假设也是了解工业建筑在转变成文化建筑后，作用于用户身上所产生的影响的关键所在。在某种程度上，我们已经意识到在探讨可使用的工业建筑时，我们将不可避免地想到文化活动，而在需要组织文化活动时，我们会想到能够唤起当地工业回忆的建筑。

作为一种说明形式，我们要做的就是查看今年的威尼斯双年展上有关Monditalia展览的评论与新闻。这次展览位于旧军械库，该地可能是工业革命之前欧洲最大的工业生产基地。大量文献将制绳厂的大型空间描述为"一个多学科文化体系的理想场所"，并且详细描述该空间是如何"集舞蹈、剧院、音乐和电影于一体的"。该场地支持这一类型的环境作为默认的新文化空间。

我们可以通过仔细阅读本章中详细说明的霍沃思·汤普森建筑师事务所开展的人人剧院项目来结束本章内容。事实上，该建筑是该事务所首个完全新建的剧院，而它看上去却像是对旧建筑的改建项目。人人剧院体现该建筑师事务所先前完成的众多剧院改造项目的一小部分，并且包含建筑师对先前工作的片段和记忆。项目最终的设计效果是一个看起来有些让人熟悉的剧院，仿佛它已然带有时间的印迹和使用过的痕迹。该建筑使用的多种材料质地粗糙，略显老化，其结构具有层次并且内部设计异常复杂。所有的这些元素使该建筑成为工业建筑再利用过程中发掘的新质量革命，并且成为我们对这些建筑保护欲的一种延续。

al amenities, we'll have better, more creative populations." Such a strategy raises all kinds of question marks as it can turn certain spontaneous and organic processes involving gentrification upside down.

An Industry Standard

If we approach this shift from industrial to cultural with an outlook of a little less gloomy, we can see a very interesting evolution at work: Industrial buildings went from being leftovers to creatively re-used sites, to become the prevailing environment for cultural institutions and eventually see its spatial and less tangible characteristics be exported to newly built cultural sites. If we have moved to an industry of culture, the buildings from the industrial era definitely have become today's industry standard. The French nineteenth century economist Alfred de Foville stated that "man makes his house and in doing so, he must put into it something of himself. However, through the passage of time, the house makes man, too." Although his statement is applied to the contemporary housing situation of his time, his hypothesis of the reciprocal influences of milieu and inhabitant is key to understanding the influences that industrial buildings, turned to cultural buildings, exercised on its users. To a certain extent we might have reached a point where cultural activities inevitably come to our minds when approaching available industrial buildings, and a building vernacular is reminiscent of the industrial when being asked to organize cultural activities.

As a form of illustration we only need to look at the reviews and press related to the Monditalia exhibition at this year's Venice biennale. This exhibition is set at the Corderie dell'Arsenale, which is possibly the single largest industrial complex in Europe prior to the industrial revolution. Numerous references describe the large space of the Corderie as "an ideal set for a multi-disciplinary cultural system" and elaborate on how the space "mobilizes dance, theatre, music and film." It simply underpins this type of environment as the new default space for culture.

We can thus end by taking a closer look at the Everyman Theatre project by Haworth Thompson architects, which is illustrated in further detail within this chapter. This building is in fact their first completely new theatre, whilst it looks like the remodelling of an old one. The Everyman Theatre embodies bits from the many theatre conversion projects they completed before, and contains fragments and memories of their previous works. The result is a theatre that seems somehow familiar, as if it already bears the traces of time and scars of use. Its multiple materials are textured and ready-aged, its structures are layered and its interiors are surprisingly complex. It makes this building altogether a new evolution of the qualities discovered through the reuse of industrial buildings and a continuation of our thirst to preserve. Tom Van Malderen

建在旧储气罐内的多功能大厅

AP Atelier

市中心附近出现了一个国家性文化地标，即带有一组独特的工业建筑的大型钢铁厂建筑群。圆柱形的储气罐直径为71.2m，经过重建，被改造成名为Gong的多功能大厅。其可移动的锥体进行了抬升，一个天篷冠于屋顶。该建筑的总高度为35.21m。混凝土和钢结构建成的储气罐能够容纳一座拥有1500个座位、用于举办音乐会和会议的大厅。贝壳形礼堂的上部坐落在22m长的管状钢柱之上。储气罐的上部和下部的洞口经过切割，并饰以玻璃，为大厅提供了宽阔的入口和大型窗口，使人们能够望见1号冶炼高炉。在主厅的下方，人们能找到一个门厅，一个小型会议室（400座）位于其楼下，艺术画廊则位于一层。Gong大厅采用了现代的视听技术。新结构是钢筋混凝土和钢的结合。

画廊和会议室上面的、发挥流线功能的塔楼和天花板都采用钢筋混凝土制成，舞台上方的平台和运用了舞台技术的结构均为钢/承重结构。舞台包含一个钢桁架梁，支撑着贝壳形的钢筋混凝土板（带有预制的铝质台阶）。原始钢结构（包括铆接的地面）的工业面貌都得以保留。

储气罐的鸣钟被提高到了顶点，以修复原有的宏观面貌。新建部分几乎不影响储气罐的原始建筑，使其仍然具有纪念性。在建筑前侧，混凝土塔楼之间建造了四个楼层。这些社会、文化和教育空间包括一个一层的画廊，一个二层的带有中央大厅的模块式会议室，以及浮于上层的大礼堂门厅。

Multipurpose Hall Fitted in Former Gasholder

A national cultural landmark near the center appears as an extensive industrial ironworks complex with a unique group of industrial buildings. This cylindrical gasholder, 71.2m in diameter has been reconstructed and transformed into the Gong Multipurpose Hall. Its movable cone was raised and a canopy was set on its roof. The total height of the building is now 35.21m. A concrete and steel construction was built into the gasholder, which hosts a 1500-seat concert and convention hall. The upper part of the shell-shaped auditorium is based on 22m long tubular steel columns. The openings were cut and glazed in both the upper and

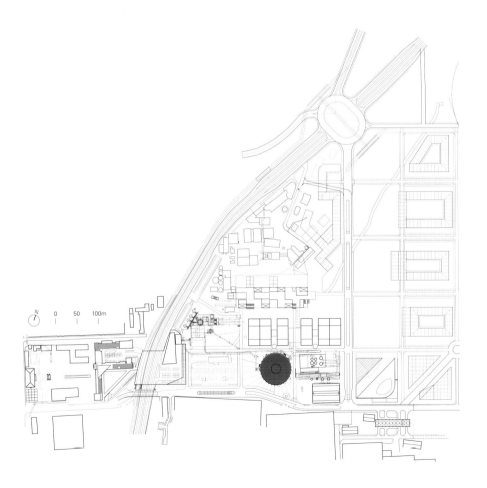

储气罐屋顶的组装，1923年
Assembling of Gasholder's roof, 1923

储气罐场地，1925年 Gasholder's grounds, 1925

维特科维策的冶金高炉，1925年
Blast Furnaces in Vitkovice, 1925

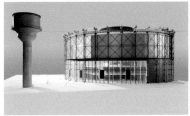

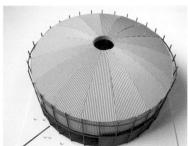

lower parts of the gasholder offering a wide entrance and a large window in the great hall with a view of Blast furnace No. 1. Below the main hall, people can find a foyer, a smaller meeting room(400 seats) on the floor below, and finally an art gallery on the first floor. Gong Hall is equipped with modern AV technology.

New constructions are a combination of ferro-concrete and steel. The circulation towers and ceiling slab above the gallery and conference sections are of ferro-concrete, and the platforms above the stage and constructions holding stage technologies are made of a steel/load-bearing construction. The stage consists of steel truss girders supporting a shell-shaped ferro-concrete slab with prefabricated auditorium steps. The industrial appearance of the original steel constructions, including the riveted floor, has been preserved.

The sung bell of the gasholder was elevated to a top position in order to restore its former majesty. The new sections barely affect the original construction of the gasholder which still has a monumental impact. In the front, four levels were constructed between concrete towers. These social, cultural and educational spaces include a gallery on the first floor, modular conference rooms with a central hall on the second floor, and the grand hall foyer that floats on the top floor.

项目名称：Multipurpose Hall Fitted in Former Gasholder
地点：Ostrava, Czech Republic
建筑师：Josef Pleskot
项目团队：Andrej kripe, Michaela Koaov, Daniel K, Milo Linhart, Zdenk Rudolf, Ji Trka
舞台设计：Jindich Smetana
舞台结构：EXCON
钢结构的生产和组装：Hutn monte
混凝土结构：RECOC
总承包商：Gemo Olomouc
建成面积：4,081m²
建筑体积：180,000m³
设计时间：2010
竣工时间：2012
摄影师：©Tomas Soucek (courtesy of the architect)

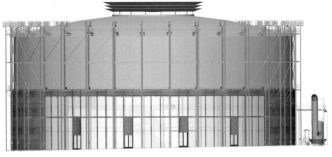
东立面 east elevation

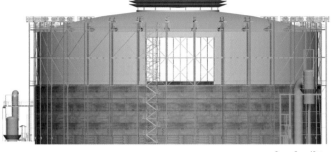
西立面 west elevation

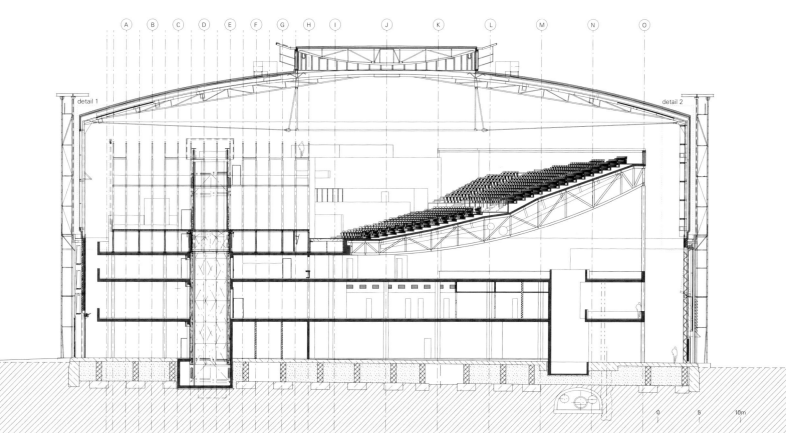

A-A' 剖面图 section A-A'

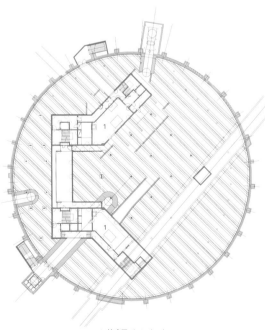

1 技术区 1. technology
地下一层 first floor below ground

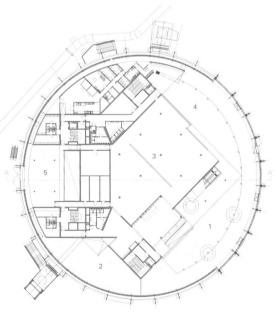

1 入口大厅 2 艺术大厅 3 艺术画廊 4 咖啡室 5 背景区
1. entrance hall 2. art hall 3. art gallery 4. café 5. background
一层 first floor

153

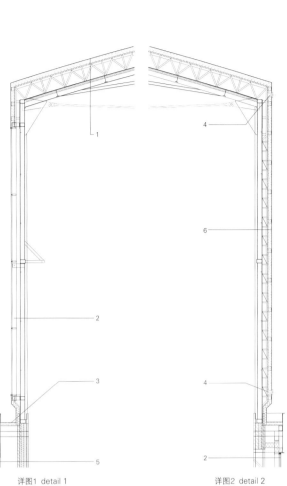
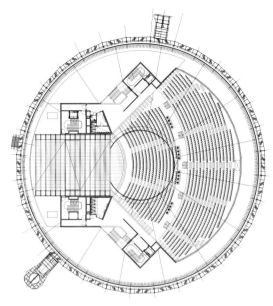
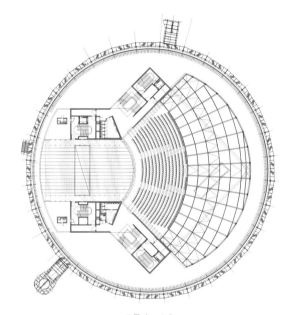
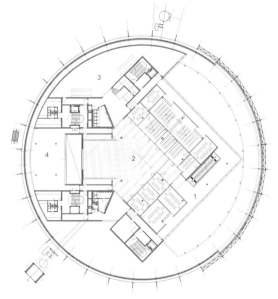
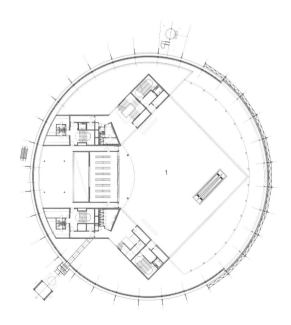

详图1 detail 1　　　详图2 detail 2

1. waterproofing membrane
   mineral wool insulation
   trapezoidal sheet
   increasing steel construction
   corrosion resistant coating
   existing steel roof
   acoustic insulation
2. aluminium facade system
   steel substructure
3. rain gutter
4. ventilated facade opening
5. existing steel structure
   air space for technology
   gas concrete wall
   mineral wool insulation
   acoustic concrete blocks
6. waterproofing membrane
   mineral wool insulation
   trapezoidal sheet
   increasing steel construction
   corrosion resistant coating
   existing steel roof

五层 fifth floor

四层 fourth floor

1 画廊　2 小厅　3 餐饮区　4 背景区
1. gallery 2. small hall 3. catering 4. background
二层 second floor

1 门厅　1. foyer
三层 third floor

C-Mine建筑对如何处理大型被遗弃的工业遗产地这一问题进行了精确的、与众不同的解答。自2005年以来,根克市一直是延伸碳带至鲁尔地区这一行动的重要合作伙伴,且正积极地重新发展过去的历史。自2010年竣工以来,老煤矿设施正稳步地成为更大区域范围内的、重要的新型文化设施节点。

列管电厂的超大规模和简单的工程方案赋予了新操作以限制主观意愿和极其直接的特色。嵌入的文化设施以原有的垂直分隔为基础——一个5m高的基座,其上是带有天窗的机械间。建筑师通过将带有两个新建的混凝土增建机构的T形砖质基座进行扩建,使一个进深长且迷人的地面层展现在人们面前;如迷宫一般的门厅设有展览空间、办公室、咖啡室、餐厅、会议室以及进出两座新剧院的通道。

在地面层,从老建筑到新建筑的扩建结构十分自然,并且在二层得以延伸。两个剧场的场地放置了一架钢琴,钢琴向上抬起,显得十分高贵。红色与白色的瓷砖表面使人们想起了原始的(浪漫的)的地面(地面是由煤矿公司自行铺设的),且自然地延伸到室外。从规模的角度来看,瓷砖给人家庭般的感觉,从看似无限的重复性角度来看,又具有工业性的特征,且对自由的、开放氛围的新老设施空间序列起到了强调的作用。

大多数原有的涡轮大厅和机械间都未经改建,在空间和功能方面,随时为其他空间的使用做准备。两座新型剧场被构思为有日光照射的机械间,从内部来看,旧砌砖建成的基础设施和钢材制成的塔楼成为新舞台的背景。因此,剧场的改造因日光和实时实景的存在而面临着挑战。在入口处,一个钢质体量过滤了广场至门厅的市民,从那里到此处有多条路线,如同一座城市一样。

C-Mine

C-Mine is a distinct and precise answer on how to deal with large-scale abandoned industrial heritage. Since 2005 the city of Genk, being an important partner in the Carbon Belt stretching as far as to the Ruhr Area, has been actively redeveloping its recent past. Since its completion in 2010 the old coalmine infrastructure is steadily becoming the new and pivotal cultural infrastructure node for the wider area.

The sheer overpowering scale and the straightforward engineering solutions of the listed power plant have rendered the new

C-Mine建筑

51N4E

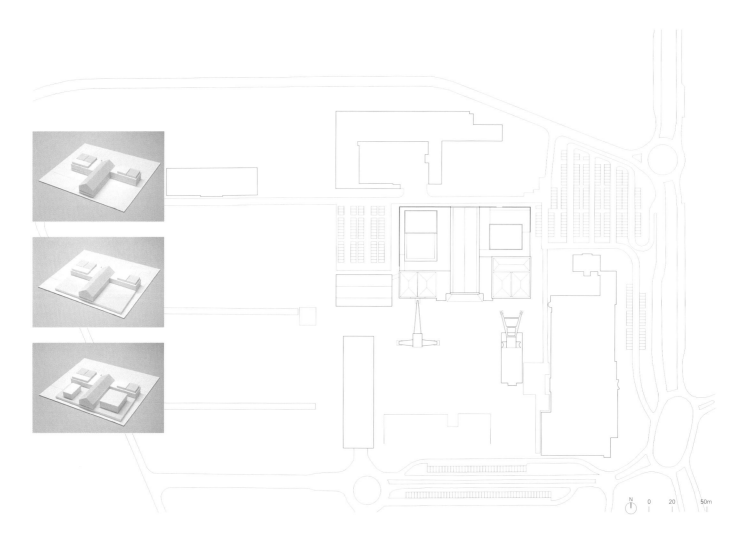

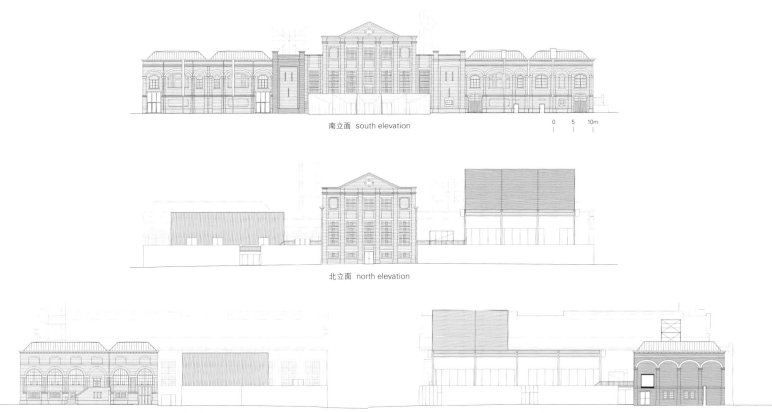

南立面 south elevation

北立面 north elevation

东立面 east elevation

西立面 west elevation

历史施工时间要点
construction historical note

- 1914
- 1916
- 1919~1920
- 1924
- 1928
- 1939
- 1947~1948
- 1949
- 1952

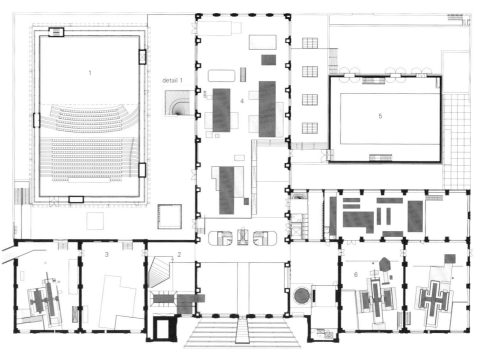

1 主礼堂 2 采矿体验馆 3 多功能空间 4 多功能大厅 5 工作室舞台 6 设计中心
1. main auditorium 2. mine experience 3. multipurpose space 4. multipurpose hall
5. studio stage 6. design centrum
二层 second floor

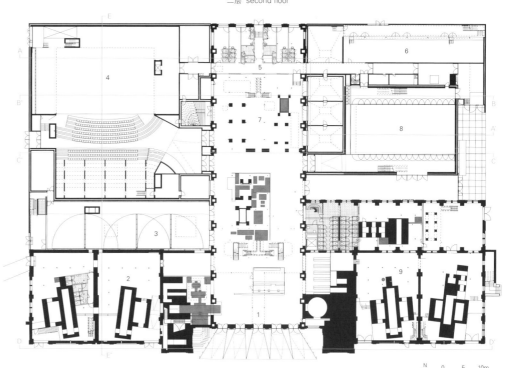

1 门厅 2 餐厅 3 展区 4 主礼堂 5 后台 6 行政区 7 酒吧的工作室 8 舞台 9 设计中心
1. foyer 2. restaurant 3. exhibition 4. main auditorium 5. backstage 6. administration
7. barstudio 8. stage 9. design centrum
一层 ground floor

operations voluntarily restricted and extremely direct. The added-on cultural infrastructure is fully based on an existing vertical division – a 5-meter high base upon which top-lit machine rooms stand. By extending the brick T-shaped base with two new concrete additions, a deep and fascinating ground level comes into being; a labyrinth-like foyer includes exhibitions spaces, offices, a cafe, a restaurant, meeting rooms and accesses to the two new theatres. The natural prolongation of the old into the new on the first level is continued on the second level; there a piano nobly arising on which the two theatre volumes stand. The red and white tiling surface, reminiscent of the original (romanticized) flooring laid out by the coalmining company itself, is literally extended to the outside. The tiles, domestic in their scale and industrial in their seemingly infinite repetition, underscore a permissive open air sequence of old and new infrastructural spaces.

Most of the existing turbine halls and machine rooms are left untouched, both spatially and programmatically, ready to be used as surplus space. The two new theatres are conceived as day lit machine rooms, from inside which the old brick infrastructure and the steel towers become the backdrop for a new stage. As a consequence, theatre conventions are challenged through the presence of daylight and real-time panorama. At the entrance a steel volume filters the public from the square into the foyer. From there the routes are multiple, like in a city.

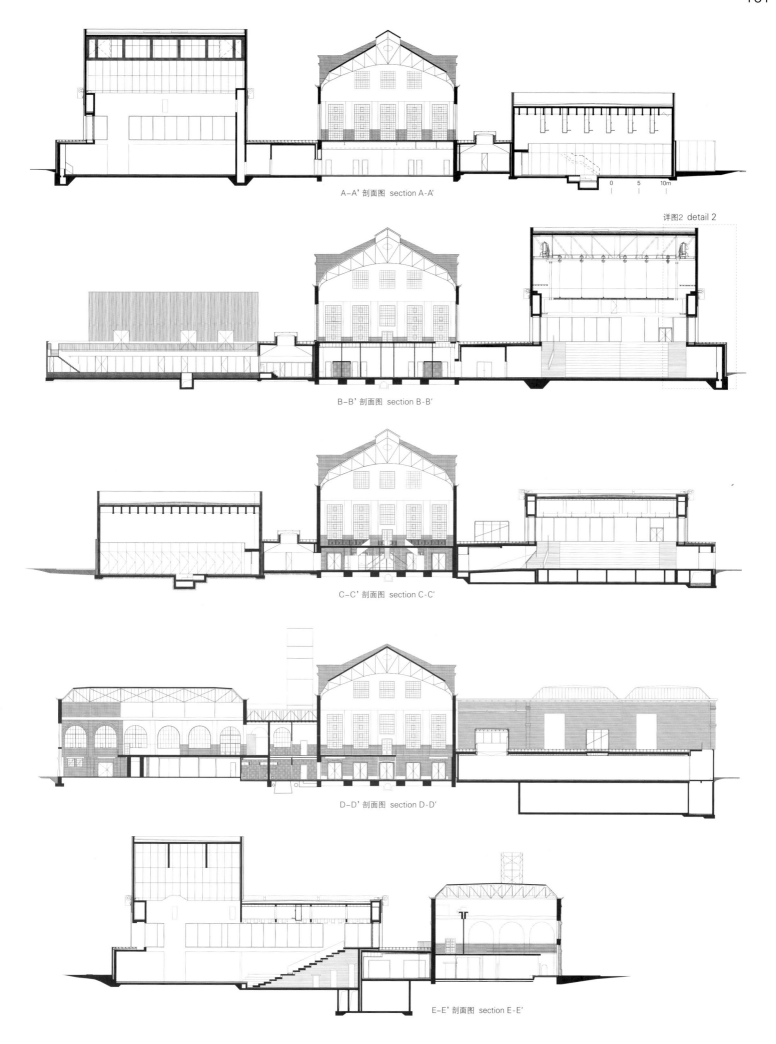

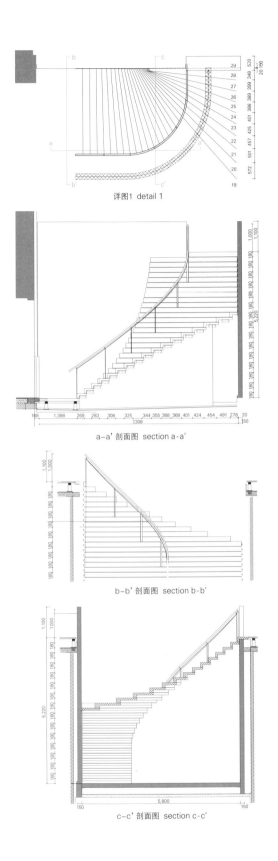

详图1 detail 1

a-a' 剖面图 section a-a'

b-b' 剖面图 section b-b'

c-c' 剖面图 section c-c'

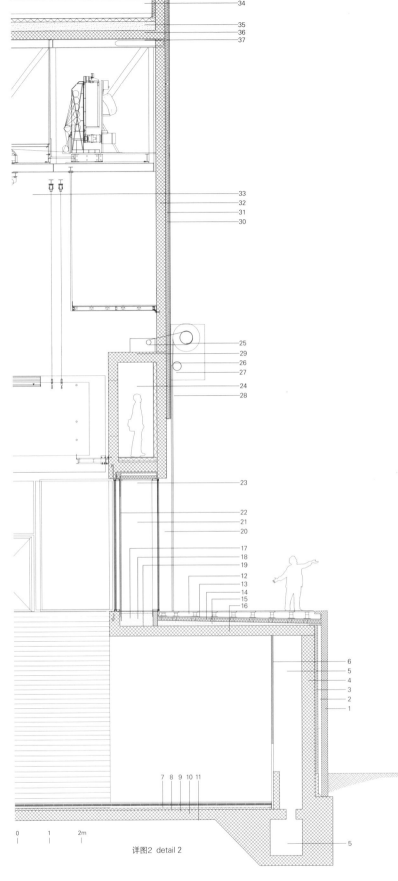

详图2 detail 2

1. wall in prestressed concrete, 180mm
2. cavity 100mm
3. thermic and acoustical insulation, mineral wool 100mm
4. load bearing wall in concrete 400mm
5. plenum
6. acoustical wall 60mm
7. fibreboard 2x22mm
8. mineral wool insulation 40mm between counter-batten 60/90
9. reinforced screed 70mm
10. thermic insulation XPS 60mm
11. load bearing concrete slab 300mm
12. ceramic tiles 15x15 (red-white)glued on concrete floor tiles 60x60
13. glued fiber cement board 2x20mm
14. compression-proof mineral wool 100mm
15. levelling concrete
16. load bearing concrete slab 300mm
17. carpet
18. screed 60mm
19. insulating lightweight concrete 400mm
20. double layered glazing, Rw>44dB
21. acoustical cavity, double glass wall of the theater space
22. darkening screen
23. suspended ceiling with acoustical insulation 220mm
24. plenum voor ventilation and access
25. axis of roll-down shutter, diam 406mm
26. opening shutters: perforated galvanized steel slats 7mm
27. pressing roll diam 273mm
28. guide rail
29. electrical motor
30. fixed shutter: perforated galvanized steel slats 7mm
31. thermic insulation mineral wool 100mm
32. load bearing concrete wall 300mm
33. stage tower
34. wall covering
35. PIR insulation 100mm
36. levelling concrete 270mm
37. load bearing concrete slab 250mm

项目名称：C-Mine Cultural Center
地点：Winterslag, Genk, Belgium
建筑师：51N4E
项目团队：Johan Anrys, Freek Persyn, Peter Swinnen, Aglaia De Mulder, Kelly Hendriks, Chris Blackbee, Joost Körver, Lu Zhang, Tine Cooreman, Aline Neirynck, Tom Baelus, Sotiria Kornaropoulou, Bob De Wispelaere, Jan Das, Philippe Nathan
合作者：Theater infrastructure_TTAS, Heritage_Bureau Monumentenzorg, Structure_BAS, Technical infrastructure_IRS, Acoustics and Physics_Daidalos-Peutz, Calculations_Probam, Main contractor_Houben
甲方：City of Genk
造价：EUR 30,000,000
功能：theater & concert hall, tourist center, design museum
用地面积：8,800m² / 总建筑面积：8,400m² / 有效楼层面积：15,500m²
受邀竞赛时间：2005 / 施工时间：2008—2010 / 竣工时间：2010
摄影师：©Stijn Bollaert

胶印厂的改造
Origin Architect

曾经的工业建筑向文化建筑的转变 Cultural Shift from bygone Industry

不同于798艺术区那些具有德国精英血统的大型工厂，蜗居在美术馆后街的北京胶印厂更像是个工业化的四合院，带着北京胡同的市井气息。几栋不同历史的工业楼房分别建于20世纪的60年代、70年代、90年代，除了层高有些许高之外并无特别之处，但却围合出了娴静的庭院。在历经半个世纪的风雨之后，原有的产业逐渐地衰落，直至完全消失，广场也逐渐荒废，面目全非。管线老化且裸露在外，而闲置的厂房被零乱地分割和出租，无序的增建使整个厂区变得拥堵，成为名副其实的大杂院。这次结合戏剧文化和功能置换的整体改造，期望为衰败的城市体注入一丝新的生机。

在尊重院落特有的历史情感的同时，延展丰富的工业文化的积累，唤醒、激发内在活力，成为改造策略的切入点。

行动首先从针对性的局部切除入手，对导致整体堵塞的增建部分进行清理：在东侧，清除临时性的私自搭建，移除杂乱的停车场，为安静的前院留有空间；而西端则拆除依附在两座主体结构之间的简陋铁皮房，形成后院；进一步清理两侧部分交通堵塞的胡同，使前后的联系畅通，让多层次的院落空间和街巷重见天日。

一条独特的空间穿行体系突破了原始结构的束缚，引导传统的院落街巷自由且立体地流动生长。有些从地面庭院迂回攀爬至屋顶花园；有些从一座房屋横向跳跃到另一座房屋，形成空中长廊；也有些直接切入室内或者地下，将阳光、空气和清新的自然引入室内空间。不再有枯燥单一的必经之路，交通动线演变成自由的立体游园小径，带动室内外空间和风景的互动。

工厂内原本无法到达且孤立的屋顶被游廊完全地激活，演变为比例适中的空中花园。每个花园的高度、景观体验、到达方式都各不相同，是立体街巷中的流动风景。随处可见的空中花园弥补了地面空间的不足，为每个内部单元都带来了亲近自然的机会，此外，用于休闲和邂逅的交流场所成为创新灵感的源泉。僵化的空间状态被打破之后，多样性的环境聚落是最适合文化创意生态发展的地方。

项目名称：Reformation of Beijing Offset Printing Factory – Cultural and Creative Park
地点：Backstreet of Art Gallery Dongcheng District, Beijing, China
建筑师：Origin Architect
总建筑师：Li Ji
设计团队：Zhang Hui, Lian Hui, Wang Jing
工程设计：Sunshine Firm
技术顾问：Ren Aidong
甲方：Beijing Dongfangdaopu Culture Assets operation management Co. Ltd.
用地面积：7,000m²
有效楼层面积：13,000m²
设计时间：2013
施工时间：2013—2014
摄影师：©Xia Zhi (courtesy of the architect)

Rebirth of the Offset Printing Factory

Different from the grand factories of elite German descent in 798, Beijing Offset Printing Factory, nestling in the backstreet of Art Gallery, is more like an industrialized courtyard with a scent of civil life in Beijing Hutongs. Built in 1960s, 1970s and 1990s respectively, the industrial buildings have different histories. They look nothing special except the higher storey height. But once upon a time, there was a secluded yard boxed in the enclosures. Unfortunately during half century of trials and hardships, the original industries declined and shut down one after another; the factories were also dilapidated day by day almost beyond all recognition. The pipe lines are now aging and bare; the idle workshops are separated and rent disorderly. The unordered addition jams the whole factory area and makes it a genuine warren. The integral reformation combining drama culture and functional replacement is in the hope of bringing a new vitality to the waste urban body.

详图1_剧院 detail 1_theater

北立面_剧院 north elevation_theater

1. brick paving
2. rain rill
3. pool, brick paving
4. return pipe
5. the water inlet pipe
6. drain-pipe
7. waterproof
8. thermal insulation
9. cast in situ method for making cement board
10. profiled steel sheet
11. 3mm thick weathering steel
12. steel I-beam
13. storm sewer
14. absorbent cotton
15. hidden lamps
16. weathering 50mm wide 3mm thick spacing of 50mm
17. anticorrosive wood floor
18. angle, connecting bar and beam
19. boom
20. heating an I-beam column groove
21. air conditioning system

a-a' 剖面图 section a-a'

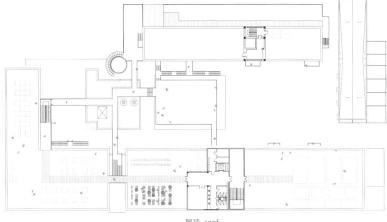

屋顶 roof

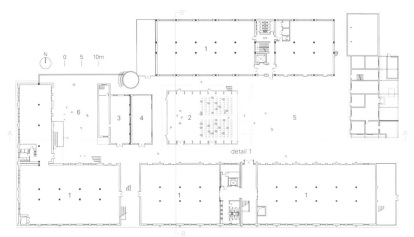

1 办公室 2 剧院 3 咖啡室 4 锅炉房 5 前院 6 后院
1. office 2. theater 3. cafe 4. boiler room 5. the front yard 6. the back yard
一层 first floor

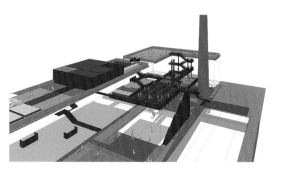

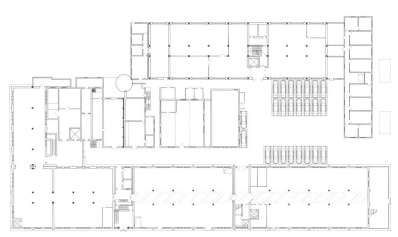

原有的一层 previous first floor

Respecting for the specific historical emotions of the courtyard, the hitting-point of the transformation strategy is to extend the accumulation of various industrial cultures, awake and arouse its inner vitality.

Starting with targeted local excision, the addition part which causes the overall blocking will be cleared: in the east, eliminate the unauthorized temporary buildings and remove the disorder parking to make room for a quiet front yard; in the west, demolish the crude tin room attached between the two major structures to form a backyard; further clean up the blocked Hutongs on both sides and link them up from back to front, thus a multi-layered courtyard space and alleys will be brought to light again.

Breaking through the shackles of original structures, a unique space traveling system is leading the free and stereoscopic flowing growth of the initial traditional yards and alleys: some climb up circuitously from the ground courtyard to the roof garden; some fly crosswise from one house to another to form an air corridor; some straightly break into the room or underground and bring the sunshine, the air and the refreshing nature inside. There will not be only route any more. Traffic moving lines will evolve into irregular three-dimensional garden trails, which brought along interaction between indoor and outdoor space and scenery.

The unreachable and isolated inactive roofs in the factory are totally activated by the verandas and becoming well-proportioned floating hanging gardens. The different heights, landscape experiences and arrival patterns of each garden form a flow of scenery in the stereoscopic alleys. Making up the shortfall of the limited ground space, the hanging gardens everywhere provide an opportunity of closing to the nature for every internal unit. Moreover, the communication place for relax and encounter will bring creative inspirations. After breaking the rigid space, the diversity of environmental settlement will grow into the most suitable place for the ecological development of cultural creativity.

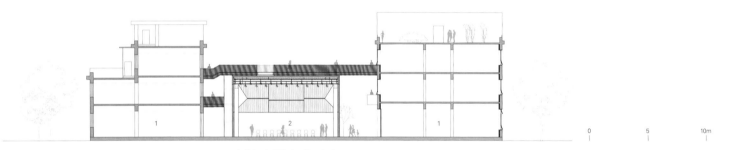

1 咖啡室　2 锅炉房　3 剧院　4 前院
1. cafe 2. boiler room 3. theater 4. the front yard
A-A' 剖面图 section A-A'

1 办公室　2 剧院　1. office 2. theater
B-B' 剖面图 section B-B'

人人剧院

Haworth Tompkins Architects

位于利物浦的人人剧院是为国际公认的制作公司而建造的新剧院，也是欧洲公开的建筑设计竞赛的获奖项目。项目作业范围包括一座拥有400个坐椅的礼堂，一处小型演出和发展空间，一个大型排练室、公共门厅、展览空间、餐饮和酒吧设施，以及配套的办公室、工作室和辅助空间。而建筑的整个立面是共同创作的一个大型公共艺术作品。其设计结合大型热能施工，一系列的自然通风系统和低能耗技术设施，使这栋复杂并且居住人口稠密的城市建筑达到BREEAM优秀评级。

人人剧院在利物浦文化中占据重要的地位。原来的剧院是对19世纪Hope Hall教堂改造得来的，它曾经是该城市的创意、欢宴和发表异议的中心（人们通常都聚集于该建筑的地下小酒馆），但随着新千年的到来，需要将该建筑整体改造以适应快速扩大的生产和参与规划。霍沃思·汤普金斯建筑师事务所的项目书在于设计一座具有先进技术和高度适应性的新剧院，它将保留旧建筑便利的可达性，映射该组织的文化价值、社区参与和本土创意，并且囊括了利物浦人民对该建筑的集体认同感。新建筑同样位于Hope街上敏感的历史文化名城的中心场地，紧邻利物浦天主教堂，周边皆为18和19世纪的列管建筑，所以达到外部公共领域的敏感度和公开性之间的平衡是该项目的一个重要的设计标准。项目书的另一个核心方面是设计一栋在建造和使用过程中都具有特殊能效性的城市公共建筑。

该建筑通过在一些半层楼周围设置公共空间的方式来利用复杂受限的场地几何形状，形成街道至礼堂之间一条不间断的弯曲散步走廊。三个楼层都安排门厅和就餐空间，包括一座位于长长的主楼层门厅最高点的、可以俯瞰街景的新酒馆。礼堂带有一处可调节的伸出式舞台空间，包括400个座位，由Hope Hall教堂的再生砖建造而成，也作为门厅的室内墙壁。该建筑由许多创意工作区组成，带有一个排练室、工作室、录音棚、俯瞰门厅的作者会面室、以及EV1——为人人剧院青年教育和社区人群准备的特殊工作室。不同的残障人群从项目伊始就监督该项目的设计。

在建筑外部，建筑师选择在墙壁和四个大型通风道上使用来自当地的红砖，使建筑呈现一种与众不同的轮廓，并且能够与周围建筑的风格相吻合。建筑的主西向立面是一个由105块可移动遮阳板组成的大型公共艺术作品，每一块板都带有一个当代利物浦居民的等身水削人像。建筑师与利物浦的摄影师丹·凯尼恩合作，使城市社区的每一部分都融入到众多的公共活动中，所以完工的建筑可以理解为城市各类人民的集体全家福。印刷商和艺术家Jake Tilson为具有代表性的红色"人人"标志的新版本设计了特殊字体，而定期与其合作的视觉艺术家Antoni Malinowski为门厅设计了一个大幅喷漆天花板块，以配合建筑内部的砌砖、碳钢、橡木、再生绿柄桑木、深色胶合板和灰白色的现浇混凝土的颜色。

The Everyman Theater

The Liverpool Everyman is a new theater which won in open European competition, for an internationally regarded producing company. The scope of work includes a 400 seat adaptable auditorium, a smaller performance and development space, a large rehearsal room, public foyers, exhibition spaces, catering and bar facilities, along with supporting offices, workshops and ancillary spaces. The entire facade is a large, collaborative work of public art. The design combines thermally massive construction with a series of natural ventilation systems and low energy technical infrastructures to achieve a BREEAM Excellent rating for this complex and densely inhabited urban building.

The Everyman holds an important place in Liverpool culture. The original theater, converted from the 19th century Hope Hall chapel, had served the city well as a centre of creativity, conviviality and dissent(often centered in its subterranean Bistro), but by the new millennium the building was in need of complete replacement to serve a rapidly expanding production and participation program. Haworth Tompkins' brief was to design a technically advanced and highly adaptable new theater that would retain the, demotic accessibility of the old building, project the organization's values of cultural inclusion, community engagement and local creativity, and encapsulate the collective identity of the people of Liverpool. The new building occupies the same sensitive, historic city centre site in Hope Street, immediately adjacent to Liverpool's Catholic cathedral and surrounded by 18th and 19th century listed buildings, so a balance of sensitivity and announcement in the external public realm was a significant design criterion. Another central aspect of the brief was to design an urban public building with exceptional energy efficiency both in construction and in use.

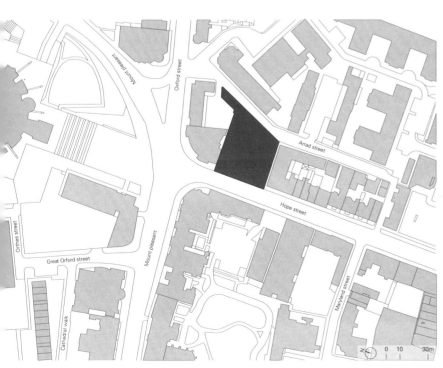

The building makes use of the complex and constrained site geometry by arranging the public spaces around a series of half levels, establishing a continuous winding promenade from street to auditorium. Foyers and catering spaces are arranged on three levels including a new Bistro, culminating in a long foyer overlooking the street. The auditorium is an adaptable thrust stage space of 400 seats, constructed from the reclaimed bricks of Hope Hall and manifesting itself as the internal walls of the foyers. The building incorporates numerous creative workspaces, with a rehearsal room, workshops, a sound studio, a Writers' Room overlooking the foyer, and EV1 – a special studio dedicated to the Young Everyman Playhouse education and community groups. A diverse disability group has monitored the design from the outset.

Externally, local red brick was selected for the walls and four large ventilation stacks, giving the building a distinct silhouette and meshing it into the surrounding architecture. The main west facing facade of the building is a large-scale public work of art consisting of 105 movable metal sunshades, each one carrying a life-sized, water-cut portrait of a contemporary Liverpool resident. Working with Liverpool photographer Dan Kenyon, the project engaged every section of the city's community in a series of public events, so that the completed building can be read as a collective family snapshot of the population in all its diversity. Typographer and artist Jake Tilson created a special font for a new version of the iconic red "Everyman" sign, whilst regular collaborating visual artist Antoni Malinowski made a large painted ceiling piece for the foyer, to complement an internal palette of brickwork, black steel, oak, reclaimed Iroko, deeply colored plywood and pale in-situ concrete.

项目名称：Everyman Theater
地点：3-11 Hope Street, Liverpool, United Kingdom
建筑师：Haworth Tompkins
室内和家具设计：Haworth Tompkins with Katy Marks
承包商：Gilbert-Ash / 项目经理：GVA Acuity
工料测量师：Gardiner & Theobald / 剧场顾问：Charcoalblue
结构工程师：Alan Baxter & Associates
服务工程师：Watermans Building Services
CDM协调员：Turner and Townsend
音效工程师：Gillieron Scott Acoustic Design
餐饮顾问：Keith Winton Design
通道设计顾问：Earnscliffe Davies Associates
合作艺术家：Antoni Malinowski
印刷商：Jake Tilson / 肖像摄影师：Dan Kenyon
甲方：Liverpool and Merseyside Theaters Trust
用地面积：1,610m² / 总建筑面积：4,690m² / 有效楼层面积：4,690m²
室外饰面：portrait panels on the facade _ 105 (etched, anodised and water jet cut 10mm aluminium sheet), bricks reclaimed from the old everyman _ 25,000
竣工时间：2013.10
摄影师：
©Hélène Binet (courtesy of the architect) - p.177[bottom], p.180~181, p.182, p.183[top], p.184
©Philip Vile (courtesy of the architect) - p.176, p.177[top], p.178~179, p.183[bottom], p.186

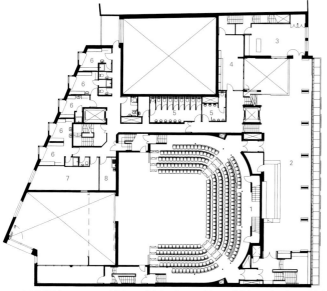

1 礼堂内的前排席 2 酒吧 3 功能间 4 作者会面室 5 卫生间 6 化妆间 7 办公室 8 洗衣房
1. auditorium stalls 2. bar 3. function room 4. writers' room 5. W.C. 6. dressing room 7. office 8. laundry
二层 second floor

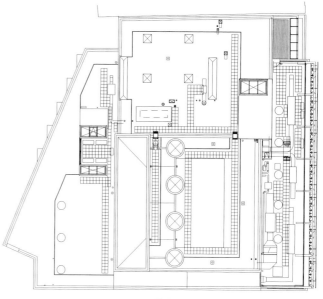

屋顶 roof

1 工作间 2 舞台 3 青年及社区工作室 4 舞台厨房 5 卫生间 6 化妆间 7 舞台管理区 8 演员休息室
9 舞台大门 10 办公室 11 青年及社区更衣室 12 垃圾房 13 商店 14 售票处 15 咖啡室
1. workshop 2. stage 3. youth & community studio 4. stage kitchen 5. W.C. 6. dressing room
7. stage management 8. actors' quiet room 9. stage door 10. office 11. youth & community changing
12. refuse store 13. store 14. box office 15. cafe
一层 first floor

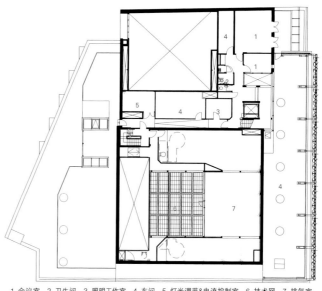

1 会议室 2 卫生间 3 照明工作室 4 车间 5 灯光调节&电流控制室 6 技术网 7 排气室
1. meeting room 2. W.C. 3. lighting workshop 4. plant 5. dimmers & amps
6. technical grid 7. extract plenum
四层 fourth floor

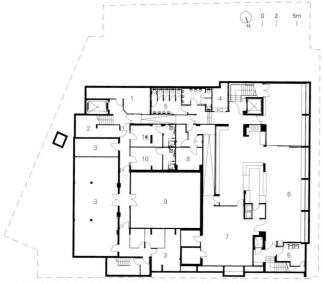

1 电气通气房 2 雨水储藏室 3 商店 4 衣帽间 5 卫生间 6 小酒吧 7 厨房 8 地窖
9 次级舞台 10 员工更衣室
1. electrical intake room 2. rainwater tank room 3. store 4. cloakroom 5. W.C. 6. bistro
7. kitchen 8. cellar 9. sub-stage 10. staff changing
地下一层 first floor below ground

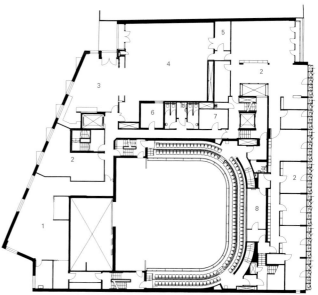

1 衣帽间 2 办公室 3 温室 4 排练室 5 商店 6 录音室 7 机房 8 控制室
1. wardrobe 2. office 3. green room 4. rehearsal room 5. store
6. sound recording room 7. server room 8. control room
三层 third floor

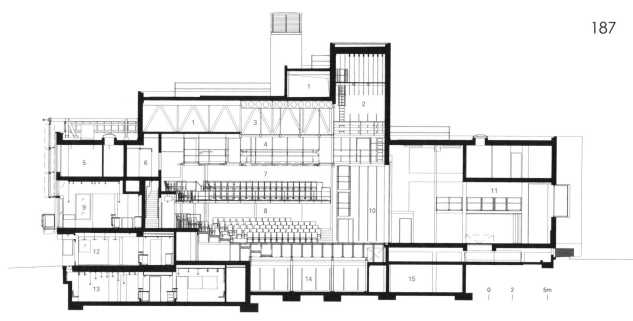

1 排气室 2 剧场塔台 3 技术网 4 技术长廊 5 办公室 6 控制室 7 环状礼堂 8 礼堂内的前排席 9 酒吧 10 舞台
11 工作间 12 咖啡室 13 小酒馆 14 次级舞台 15 商店
1. extract plenum 2. fly tower 3. technical grid 4. technical gallery 5. office 6. control room 7. auditorium circle
8. auditorium stalls 9. bar 10. stage 11. workshop 12. cafe 13. bistro 14. sub-stage 15. store

A-A' 剖面图 section A-A'

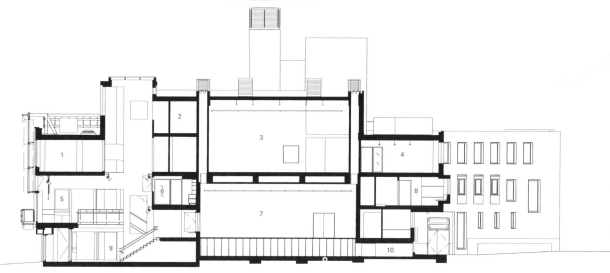

1 办公室 2 洗手间 3 排练室 4 温室 5 酒吧 6 作者会面室 7 青年&社区工作室 8 更衣室 9 咖啡室 10 进气室
1. office 2. W.C. 3. rehearsal room 4. green room 5. bar 6. writers' room 7. youth & community studio
8. dressing room 9. cafe 10. air supply plenum

B-B' 剖面图 section B-B'

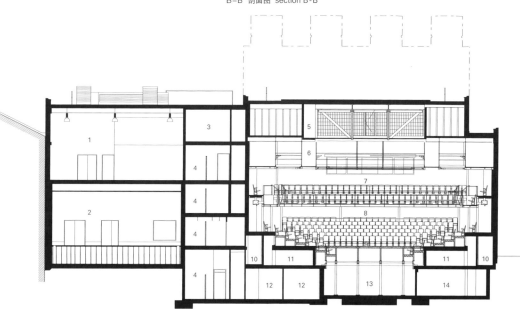

1 排练室 2 青年&社区工作室 3 车间 4 洗手间 5 技术网 6 技术长廊 7 环状礼堂 8 礼堂内的前排席
9 舞台 10 走廊 11 进气室 12 员工更衣室 13 地窖 14 商店
1. rehearsal room 2. youth & community studio 3. plant 4. W.C. 5. technical grid 6. technical gallery 7. auditorium circle
8. auditorium stalls 9. stage 10. corridor 11. air supply plenum 12. staff changing 13. cellar 14. store

C-C' 剖面图 section C-C'

>>72
MAYU architects
Was founded in 1999 under the name of Malone Chang Architects and it was changed into Malone Chang and Yu-lin Chen Architects. Recognizes the importance of teamwork in the creation of architecture, while maintaining the individuality and artistry of each project. Organizational logic of physical buildings, holistic experience of spaces, and the changing characteristics of materials are priorities of the studio. Malone Chang graduated in 1979 at the architectural department of National Cheng-Kung University(NCKU). Received a master's degree from the University of Illinois in 1983 and a doctorate of Philosophy in Architecture from NCKU in 1999. Is currently teaching at NCKU and Shu-Te University. Yu-lin Chen received a M.Arch with distinction from Harvard University Graduate School of Design in 2002 and founded Yu-lin Chen Architects in 2006. Has been a visiting design critic of NCKU since 2006.

de Architekten Cie
Branimir Medić and Pero Puljiz graduated from the faculty of architecture of Zagreb University, Croatia, in 1987 and received a master of excellence in architecture from the Berlage Institute in Amsterdam, 1992. They established Medić + Puljiz with an alliance to de Architekten Cie in 1995. After several years(1998), they joined de Architekten Cie as partners. Branimir Medić has lectured at the Academies of Architecture in Amsterdam, Rotterdam and Tilburg, as well as at Delft University of Technology. Pero Puljiz has lectured at the various Dutch academies of architecture and the University of Zagreb. Since 2009, he has been a member of the Dutch Green Building Council.

>>154
51N4E
Is a Brussels-based international practice founded in 1998. Is led by two partners; Johan Anrys, Freek Persyn and is at present 28 people strong. Johan Anrys and Freek Persyn were both born in 1974 in Belgium and studied architecture in Brussels and Dublin, graduating in 1997. Aspires to contribute, through means of design, to social and urban transformation. Concerns itself with matters of architectural design, concept development and strategic spatial transformations.

>>112
Lahdelma & Mahlamäki Architects
Ia a Helsinki-based architecture firm founded in 1997 by Mahlamäki and Lahdelma. The Finnish architect and professor Rainer Mahlamäki was born in 1956. Studied architecture at the Tampere University of Technology, and was awarded the Master of Science in Architecture in 1987. In 2007, he was appointed President of the Finnish Association of

>>84
Herzog & de Meuron
Was established in Basel, 1978. Has been operated by senior partners; Christine Binswanger, Ascan Mergenthaler and Stefan Marbach, with founding partners Pierre de Meuron and Jacques Herzog. Has designed a wide range of projects from the small scale of a private home to the large scale of urban design. While many of their projects are highly recognized public facilities, such as their stadiums and museums, they have also completed several distinguished private projects including apartment buildings, offices and factories. The practice has been awarded numerous prizes including "The Pritzker Architecture Prize" in 2001, the "RIBA Royal Gold Medal" in 2007.

>>62
Rojkind Arquitectos
Is a Mexico City based architecture firm practicing internationally focusing on tactical and experiential innovation. Uses design thinking to cut across strategic fields looking to maximize project potential while maintaining attainability. Their multinational team works hand in hand with clients and collaboratively with experts in different fields to attain the necessary knowledge and to gain value and also to establish methods in service of the project and its areas of influence. Received worldwide recognition for its award winning projects including 2005 Architectural Record Top 10 Design Vanguard. Has been runn by a partner, Gerardo Salinas[left] and a founding partner Michel Rojkind[right].

>>138
AP Atelier
Josef Pleskot was born in 1952 in Písek, Czech Republic. Graduated from the faculty of architecture of the Czech Technical University in Prague(CVUT) in 1979. In 1995, he won the Czech Society of Architects' GRAND PRIX and became a member of S.V.U. Mánes, a society of artists. Participated in the Venice Biennial Art Exhibition in 1991 and 1996. Works together with 12 colleagues in his team-AP Atelier. Regularly exhibits at its annual expositions.

>>100
Trahan Architects
Brad McWhirter, an associate of Trahan Architects, graduated from Louisiana State University in 2002 with a B.Arch. His expertise has contributed to Trahan Architects becoming known for its ability to create performance space that draws people and communities together for unforgettable experiences and for designing facilities that give an iconic presence to cities. Victor Trahan is a president of Trahan Architects. Is a 1983 graduate of Louisiana State University and is a NCARB accredited architect. Has been recognized both nationally and internationally for innovative design and creative use of materials. Was elected to the AIA College of Fellows in 2006 at the age of 45. Is currently teaching a graduate level design studio at MIT School of Architecture and Planning in Cambridge, Massachusetts.

Douglas Murphy
Studied architecture at the Glasgow School of Art and the Royal College of Art, completing his studies in 2008. As a critic and historian, he is the author of The Architecture of Failure(Zero Books, 2009), on the legacy of 19th century iron and glass architecture, and the forthcoming Last Futures (Verso, 2015), on dreams of technology and nature in the 1960s and 1970s. Is also an architecture correspondent for Icon Magazine, and writes regularly for a wide range of publications on architecture and culture.

Tom Van Malderen
His activities stretch from the traditional architectural practice to the field of architectural theory which he explores through writing, installations and lectures. After obtaining a master in Architecture at LUCA, Brussels(1997) he worked for Atelier Lucien Kroll in Belgium and in different positions at architecture project, both in the UK and Malta. Lectured at the University of Aix-en-Province in France and the Canterbury University College of Creative Arts in the UK. Contributes to several magazines and publications, and sits on the board of the NGO Kinemastik for the promotion of short film.

>>38
Estudio Carme Pinós
International Fellowship of the RIBA and Honorary Member of the AIA. Graduated from the ETSAB in Barcelona in 1979 and set up her own studio in 1991. Has received the First Prize at the 9th Spanish Biennial of Architecture and Urban Planning and the National Award of Spanish Architecture. Has taught at the University of Illinois at Urbana-Champaign, Arts Academy of Düsseldorf, Columbia University in New York, Federal Institute of Technology Lausanne, ETSAB in Barcelona and at Harvard University Graduate School of Design.

Haworth Tompkins Architects

Was founded in 1991 by Graham Haworth and Steve Tompkins. Has an international reputation for awarding winning theater design. Was part of the Gold Award UK winning team at the Prague Quadrennial and was chosen to exhibit work at the 2012 Venice Biennale. Steve Tompkins is a contributor to Theatrum Mundi, a research forum led by Professor Richard Sennett, which brings architects and town planners together with performing and visual artists to re-imagine the public spaces of 21st century cities.

>>166

Origin Architect

Was founded to advocate starting from the reality of the present complex changes, trying to look for the path of integration of globalization and regional culture, integration of individual space and the overall environment, integration of city and nature, and integration of future demand and historical Inheritance, through careful observation and thinking of the local environment. Li Ji is the founder and chief architect of ORIGIN ARCHITECT. Graduated from the Tsinghua University School of Architecture in 1994. Is a Class 1 Registered Architect of PRC (People's Republic of China) and a member of the Hong Kong Institute of Architects(HKIA).

>>50

Neutelings Riedijk Architecten

Is an architecture firm based in Rotterdam, the Netherlands, founded in 1987 by Willem-Jan Neutelings and Michiel Riedijk. Offers a strong commitment to design excellence, realizing high quality architecture by developing powerful and innovative concepts into clear built form. Over the past twenty-five years, they have specialized in the design and realization of complex projects for public and cultural buildings, such as museums, theaters, concert halls, city halls and libraries. The office has received awards such as the Golden Pyramid and the Rotterdam Maaskant Prize.

>>124

Barbosa & Guimarães Arquitectos

José António Barbosa and Pedro Guimarães graduated from the faculty of architecture of the University of Porto in 1993 and established Barbosa & Guimarães Arquitectos in 1994. They develop projects in various fields of architecture, ranging from public procurement contracts, as well as through private contracts. The valorization of the place spirit through a simple and beautiful gesture defines the atelier attitude in search of a own path intended to be consistent and contextual. Their projects are developed by valuing the details of all building elements, providing a constant search for new technical solutions suitable to the specificity of each project.

>>24

De Zwarte Hond

General director and Senior architect, Jurjen van der Meer[left] was born in Zwaagwesteinde, the Netherlands in 1953. Graduated from the faculty of architecture at the University of Technology Delft in 1981 and founded De Zwarte Hond in 1985. Tjeerd Jellema[right] was born in Grootegast, the Netherlands in 1971. Studied Architectural design at Art academy ABK Minerva, Hanze University Groningen and Academy of Architecture. Has been working as architectural designer De Zwarte Hond since 1997.

C3, Issue 2014.9

All Rights Reserved. Authorized translation from the Korean-English language edition published by C3 Publishing Co., Seoul.

© 2014 大连理工大学出版社
著作权合同登记06-2014年第174号
版权所有·侵权必究

图书在版编目(CIP)数据

建筑的文化意象 / 韩国C3出版公社编；时真妹等译.
—大连：大连理工大学出版社，2014.10
（C3建筑立场系列丛书）
书名原文：C3 Cultural Image of Architecture
ISBN 978-7-5611-9576-5

Ⅰ.①建… Ⅱ.①韩… ②时… Ⅲ.①建筑－文化
Ⅳ.①TU-8

中国版本图书馆CIP数据核字(2014)第236614号

出版发行：大连理工大学出版社
　　　　　（地址：大连市软件园路80号　邮编：116023）
印　　刷：上海锦良印刷厂
幅面尺寸：225mm×300mm
印　　张：12
出版时间：2014年10月第1版
印刷时间：2014年10月第1次印刷
出 版 人：金英伟
统　　筹：房　磊
责任编辑：张昕焱
封面设计：王志峰
责任校对：高　文

书　　号：978-7-5611-9576-5
定　　价：228.00元

发　行：0411-84708842
传　真：0411-84701466
E-mail：dutp@dutp.cn
URL：http://www.dutp.cn